Editorial Manager	Chester Fisher
Senior Editor	Lynne Sabel
Editor	John Rowlstone
Assistant Editor	Bridget Daly
Designers	QED (Alastair Campbell and Edward Kinsey)
Series Consultant	Keith Lye
Consultants *(The Universe)*	John Ebdon
	Iain Nicolson
Consultant *(The Earth)*	Keith Lye
Production	Penny Kitchenham
Picture Research	Jenny Golden

© Macdonald Educational Ltd. 1978
First published 1978
Macdonald Educational Ltd.
Holywell House
Worship Street
London EC2A 2EN

2081/3200
ISBN 0 356 05761 5

Designed and created in
Great Britain

Printed and bound by
New Interlitho, Italy

WORLD OF KNOWLEDGE

The Universe and the Earth

Neil Ardley, Ian Ridpath and Peter Harben

Macdonald

Contents

The Universe

The Earth

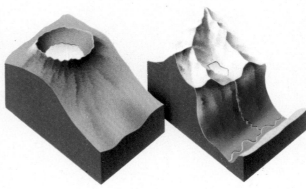

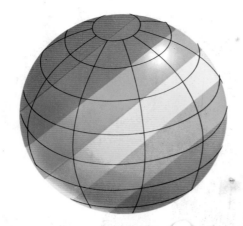

World of Knowledge

This book breaks new ground in the method it uses to present information to the reader. The unique page design combines narrative with an alphabetical reference section and it uses colourful photographs, diagrams and illustrations to provide an instant and detailed understanding of the book's theme. The main body of information is presented in a series of chapters that cover, in depth, the subject of this book. At the bottom of each page is a reference section which gives, in alphabetical order, concise articles which define, or enlarge on, the topics discussed in the chapter. Throughout the book, the use of SMALL CAPITALS in the text directs the reader to further information that is printed in the reference section. The same method is used to cross-reference entries within each reference section. Finally, there is a comprehensive index at the end of the book that will help the reader find information in the text, illustrations and reference sections. The quality of the text, and the originality of its presentation, ensure that this book can be read both for enjoyment and for the most up-to-date information on the subject.

The Universe

Neil Ardley and Ian Ridpath

Introduction

Since the start of the Space Age on 4 October, 1957, our knowledge of the Solar System and beyond has increased enormously and **The Universe** is an up-to-date survey of our present knowledge and the latest theories based upon it. The scope of coverage is wide-ranging. **The Universe** includes a historical survey of astronomy from early times, an account of the development of telescopes and modern space technology, and information on the pioneers, scientists and astronauts who have contributed so much to our present understanding. There is also a description of the Universe and full accounts of our present knowledge about each of the heavenly bodies in the Solar System.

The Universe is also forward-looking. It describes various perplexing mysteries of space which have yet to be solved and indicates the kinds of development we can expect in the years ahead.

Early astronomers once had the status of priests, but astronomy gradually developed into an important science. Although many of the mysteries of the Universe have been solved in the last 400 years, many problems still puzzle astronomers today.

Understanding the Universe

the golden fleece. live...at on

Man's exploration of the heavens, whether with the naked eye, telescope or spacecraft, has enabled us to gain some idea of our relative importance. We are the most advanced life-form on the Earth, but our observations have revealed that our world is but a tiny and insignificant corner of the Universe. We have not found life beyond the Earth and our continuing search of the Universe may show that other regions are incapable of supporting life as we know it.

However, progress in cosmology – the study of the nature of the Universe – has long been impeded by ideas that everything revolves around us and that we occupy a position at the centre of the Universe. As a result, the evolution of our knowledge of the Universe has taken place in fits and starts. Man began to investigate the

Above: The telescope reveals millions of stars in the sky, many of them like our Sun and probably possessing planets. This view of a small part of the Milky Way in the constellation Cygnus (the Swan), also shows a nebula, a cloud of glowing gas in which stars are probably being formed. It is called the North American nebula because it resembles the continent in its shape.

Universe about 2,500 years ago, but progress has only taken place in bursts for about a quarter of this time. If progress had not been interrupted by speculation at odds with the evidence of our own eyes, how much more would we know today?

The ancient world

People in every civilization have looked up at the sky and placed their own interpretations on the sights there. In some mythologies, the Sun and Moon are benevolent gods or goddesses; the Sun gives warmth by day and the light of the Moon may banish the terrors of the night. To ancient peoples, who saw the outlines of animals in the patterns made by the stars, an eclipse was as if a serpent or some other beast had swallowed their deities and had to be slaughtered by some other

Reference

A **Aberration of starlight,** discovered by James BRADLEY, is an effect of the Earth's motion around the Sun, which causes slight changes in the measured positions of stars. As an analogy, imagine driving through rain. Although the rain may be falling vertically, the car's forward motion makes the raindrops seem to come from in front. In the same way, the Earth's motion across incoming

light from stars introduces errors in the measured direction of the stars.
Anaximander (c.610-c.546 BC) was a Greek scientist often called 'the Father of Astronomy'. He believed that the Earth was shaped like a cylinder, around which the heavens rotated daily.
Andromeda galaxy is a spiral galaxy of stars 2·2 million light years away; the farthest object visible to the naked eye. On clear nights it can be seen as a fuzzy patch near the centre of the constellation Andromeda. It is similar to our own MILKY WAY.

Aristarchus (3rd century BC) was a Greek scientist who believed that the Sun was the centre of the Universe. He was the first to

Andromeda galaxy

calculate the relative sizes of the Sun, Earth, and Moon. He found that the Moon was over 30% of the diameter of the Earth (very close to the true value of 0·27), and estimated that the Sun was 7 times the Earth's diameter. This is about 15 times too small.
Aristotle (384-322 BC) was a Greek scientist who believed that the Sun, Moon, and planets went around the Earth on the surfaces of a complicated series of heavenly spheres. He knew that the Earth and Moon were spherical and that the

Moon shone by reflecting sunlight, but he did not believe that the Earth moved in space or spun on its axis. His authority was used to refute all such ideas until the time of COPERNICUS.
Astrolabe was an ancient astronomical instrument. By pointing a sighting rod at the stars, their altitude and the local time could be read off.

B **Background radiation** is a slight warmth throughout space, believed to be the heat left over from the BIG BANG explosion that started the Universe expand-

Ptolemy's theory

Copernicus's theory

Above left: Most people in ancient times believed that the Earth was the centre of the Universe and that the Sun and the other 5 planets then known revolved around the Earth. This theory was propounded by Ptolemy.

Above right: Copernicus, who lived in the 1400s followed a Greek, Aristarchus, in saying that the Earth moves around the Sun together with the other planets, though neither had any proof of this.

Below: The ancient Egyptians, like other ancient peoples, worshipped gods in the sky. The sky-goddess Nut is here shown surrounded by signs of the zodiac on a wooden coffin.

Sun
Mercury
Venus
Earth
Moon
Mars
Jupiter
Saturn
Uranus
Neptune
Pluto

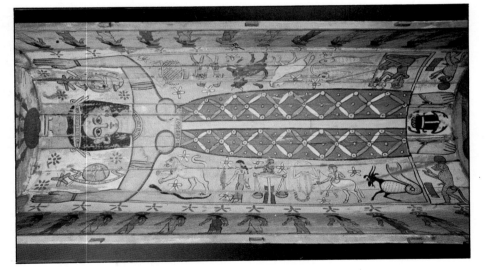

god in the sky to enable the Sun or Moon to live again. Consequently, complex mythologies developed about whole families of gods residing in the heavens. Various explanations also grew of the nature of the Earth itself and how it began. In many such myths, the world emerges from primeval waters and floats in them.

However, together with these legends came observations of the movements through the sky of the Sun, Moon, planets and stars. All ancient peoples must have realized that the seasons come and go in time with the changes of the position of the Sun in the sky. They evolved CALENDARS to predict the passage of the seasons and gradually refined them with experience. Thus, the Babylonian calendar of 354 days to the year gave way to the Egyptian calendar of 365 days, and this in

---

ing. Although the big bang must have created a temperature of many millions of degrees, as the Universe has expanded its temperature has fallen to the current figure of 2·7° Kelvin (2·7 degrees above absolute zero, the coldest temperature possible). The existence of such a background warmth, predicted in 1948 by George GAMOW, was discovered by American radio astronomers in 1965.

Barred spiral galaxy is a galaxy in which the stars and gas at its centre are arranged into a long, straight bar. At each end of the bar begins a curving arm of stars, as in a normal spiral galaxy.

Big Bang is the giant explosion which is believed to

Stephan's quintet in Pegasus

mark the origin of the Universe as we know it. Before then, all the material of the Universe was compressed into a superdense state. Gas thrown out from the big bang condensed to form galaxies of stars, which continue to move outwards as the Universe expands. According to the latest measurments, the big bang occurred about 20,000 million years ago.

Bondi, Sir Hermann (1919–) is a British astronomer who, with Thomas GOLD and Fred HOYLE, put forward the STEADY STATE THEORY, which denies that the Universe began with a BIG BANG. From 1967–71 Bondi was director-general of the ESRO (*see page 52*).

Bradley, James (1693–

Sir Hermann Bondi

1762) was an English astronomer who in 1728 discovered the effect known as the ABERRATION OF STARLIGHT, which was the first direct

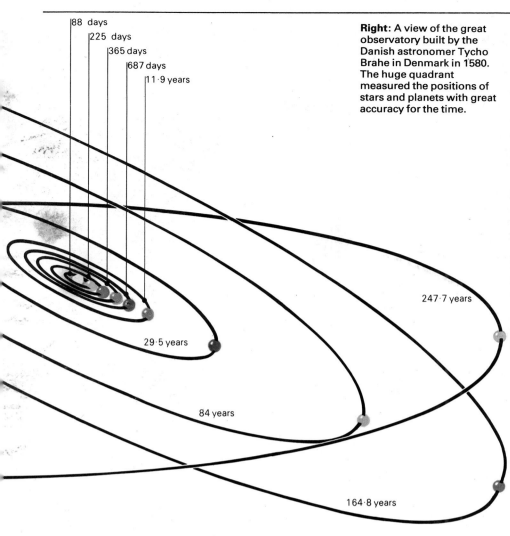

88 days
225 days
365 days
687 days
11·9 years

247·7 years

29·5 years

84 years

164·8 years

Right: A view of the great observatory built by the Danish astronomer Tycho Brahe in Denmark in 1580. The huge quadrant measured the positions of stars and planets with great accuracy for the time.

turn to the Roman system of 365.25 days.

The dawn of astronomy

Ancient civilizations developed simple instruments such as the ASTROLABE to measure the positions of the stars and planets. It is possible that monuments, such as Stonehenge, were ancient observatories where precise observations of the heavens could be made. In this way, good calendars were produced and it may have been possible to predict eclipses of the Sun and Moon. Such a feat would have appeared magical to superstitious people, and astronomers came to be held in high regard.

The first astronomer of whom much is known was Thales, who lived in Greece. He predicted the eclipse of the Sun that took place in 585 BC,

observational proof that the Earth orbits the Sun. From the amount of aberration he calculated the speed of light at 295,000 km/sec, only slightly slower than the true value. He became Astronomer Royal in 1742.
Brahe, Tycho (1546–1601) was a Danish astronomer, considered to have been the greatest observer of pretelescopic days. Using simple sighting instruments, Brahe measured the positions of the planets with greater accuracy than anyone before, enabling his assistant, Johannes KEPLER,

to work out the laws of planetary motion.

C **Calendar** is a timetable based on the movements on the Earth. The calendar was the first practical application of astronomy, and a vital one for farmers who needed to know the exact times for planting and reaping crops. Each year is one orbit of the Earth around the Sun, and each day is one rotation of the Earth on its axis. There are approximately 365·25 days in a year, so that extra days have to be added in

leap years to keep the calendar in step with the planetary motions. Our modern arrangement of leap years every four years was intro-

Shepherd's calendar

duced in 1582 by Pope Gregory XIII (Britain and the US did not change until 1752). A month is approximately the time the Moon takes to complete its phases; hours, minutes, and seconds are purely man-made divisions of time.
Copernicus, Nicolaus (1473–1543) was a Polish astronomer who proposed that the Earth was not the centre of the Universe, as previously assumed, but instead orbited the Sun like the other planets. This bold scheme was presented in his book *On the Revolutions of*

the Celestial Spheres, published in the year of his death. His scheme made it easier to account for the observed planetary move-

Nicolaus Copernicus

Right: A star map published in about 1660 shows the constellations as elaborate figures. Lines of celestial latitude and longitude mark the positions of the stars in the sky.

Below: Stonehenge is one of several groups of standing stones that might have been used as observatories in ancient times. By observing how the motions of the Sun and Moon line up with the stones, one is able to predict eclipses and work out an accurate calendar. However, the mathematics involved are not simple, and it is possible that these ancient sites were in fact temples and not observatories at all.

stars and planets, and realized that the motions of the planets against the background of stars are not in perfect circles around the Earth. But instead of adopting Aristarchus's views, which were in fact correct, he developed the ideas of Aristotle and Plato and said that the planets move in EPICYCLES – circles within circles – around the Earth.

The Dark Ages

Greek knowledge did not completely disappear in the 'Dark Ages' following the fall of the Greek and Roman civilizations. The Arabs kept it alive, drawing principally from the *Almagest*, a summary of Greek astronomy by PTOLEMY, who adopted the system of epicycles to explain the nature of the Universe. However, the Arabs did

ments. His theory was supported by the observations of GALILEO and confirmed by the calculations of Johannes KEPLER.

Cosmic rays are very fast atomic particles moving through space at nearly the speed of light. The Earth is shielded from cosmic rays by its magnetic field and atmosphere, but cosmic rays present a danger to astronauts in space.

D **Döppler effect** is the change in wavelength of radiation received from a moving object. The effect is most familiar in sound, when a vehicle's horn or siren appears to change in pitch as it passes the listener. The effect also applies to light waves, lengthening them (making them redder) if the object is receding (the RED SHIFT) and shortening them (making them bluer) if the object is approaching. The effect is named after the Austrian physicist, C. J. Döppler, who drew attention to it in 1842.

Dreyer, Johan Ludwig Emil (1852–1926) was a Danish astronomer who compiled a major catalogue of nearly 8,000 star clusters and nebulae, called the *New General Catalogue* (NGC). Many objects, particularly galaxies, are still referred to by their NGC numbers.

E **Elliptical galaxy** is a galaxy of stars with a shape ranging from a sphere to a flattened disk. Most elliptical galaxies look like a rugby ball or American football in cross-section. Giant elliptical galaxies are the largest galaxies known, containing up to 10 million million stars. At the other end of the scale, dwarf elliptical galaxies may contain only a million stars.

Epicycle is a device once used by astronomers to help explain the observed motions of the planets. The epicycle was a small circle traced out by a planet as it moved along a much larger orbit known as the deferent. Such systems of circles were done away with when Johannes KEPLER discovered that planets orbit the Sun in ellipses.

Eratosthenes (c.276–c.194 BC) was a Greek astronomer who made the first accurate measurement of the size of the Earth. He noted the different altitude of the Sun in the sky as seen on a given date from two places of known distance apart on a north-

Galaxy in Perseus

not try to develop the volume of astronomical knowledge that they were guarding. Before progress was again possible, astronomy had to be reintroduced to Europe via Spain during the Middle Ages.

Renaissance

The revival of knowledge – the Renaissance – that took place in the late Middle Ages gave rise to a marvellous flowering of the arts and sciences. In astronomy however, the rebirth was difficult and painful. The views of Plato and Aristotle had become so accepted that few people were willing to listen to other ideas. The Polish astronomer Nicolaus COPERNICUS saw that the motions of the planets could be explained by assuming that the Earth moves around the Sun, together with the other planets. But Copernicus was reluctant to publish his views. He feared ridicule as he could not actually *prove* that he was correct. His disciple Georg Rheticus convinced Copernicus that he must get his ideas across to people, and his book was finally published in 1543, the year of his death. Copernicus's book on the revolutions of the planets brought the word 'revolutionary' into being.

Copernicus introduced the concept of infinity into astronomy, implicitly banishing God from the heavens. This was heresy to some. The greatest observational astronomer of the century, Tycho BRAHE, held to a geocentric (Earth-centred) view of the Universe, maintaining that the other planets move around the Sun as it moves around the Earth! Absurd as this seems,

Above: Galileo demonstrates the telescope to the Doge of Venice, telling him of the marvellous sights to be seen in the heavens. Modern astronomy dates from 1609, the year in which Galileo began to use the telescope to observe the Sun, Moon, planets and stars and Kepler explained the laws of planetary movement.
Below: The Bayeux tapestry, which contains scenes from the invasion of England by the Normans in 1066, portrays the arrival of a comet as an omen of disaster for King Harold. The comet was in fact Halley's comet.

the theory could not be disproved. Religious divisions hardened the attitude of the Roman Catholic church to this geocentric view and in Italy, GALILEO was persecuted in 1633 for spreading the heliocentric (Sun-centred) views of Copernicus. Meanwhile, Johannes KEPLER had shown from Brahe's observations that all planetary motions could be exactly accounted for if the planets move in ellipses around the Sun and not in circles.

Progress in astronomy now moved to England. The theoretical work that had begun so long before in ancient Greece came to a triumphant climax with the discovery by Isaac NEWTON that gravity holds the planets and moons in their orbits. His theory was first published in 1687. Emphasis in astronomy shifted to observation, following the pioneering discoveries of Galileo with the newly-invented telescope. In 1728, James BRADLEY obtained final proof that the Earth moves by observing the ABERRATION OF STARLIGHT. Sir William HERSCHEL discovered a new planet, Uranus, in 1781. Its motion suggested the presence of another planet, leading to the discoveries of Neptune in 1846 and Pluto in 1930.

Beyond the Solar System
With powerful telescopes, astronomers could explore far out into space. They discovered many different kinds of stars and, far beyond the Solar System, strange glowing clouds known as NEBULAE. Charles MESSIER and Johan DREYER

south line. He calculated from these observations that the Earth must be 13,000 km in diameter, almost exactly the correct figure.

F Flamsteed, John (1646–1719) was the first Astronomer Royal. He became director of the Royal Observatory at Greenwich, England, on its opening in 1676. His task was to prepare tables of star positions and the Moon's motion to help navigators at sea. His catalogue of nearly 3,000 stars was published after his death by his assistant.

G Galaxy is an aggregation of millions of stars, bound together by gravity. Some galaxies are spiral in shape, like our own MILKY

Galaxy in Triangulum

WAY and the ANDROMEDA GALAXY, while others are dense blobs known as ELLIPTICAL GALAXIES. There may be a million million galaxies (with millions of stars in each) visible through the largest telescopes.
Galileo Galilei (1564-1642) was an Italian scientist who created a revolution in astronomy with his pioneering observations of the heavens. In 1609 Galileo heard of the invention of the telescope and built one for himself, with which he discovered craters on the Moon, saw that Venus showed phases

as it orbited the Sun, and found that Jupiter had four tiny moons that orbited it like the planets moving around the Sun.
Gamow, George (1904-68) was an American astronomer who supported the BIG BANG theory of the Universe. Gamow calculated that about 10% of the material in the Universe should be helium gas, made from hydrogen during the big bang; observations have confirmed this prediction. He also predicted the existence of a slight warmth in the Universe, left over from

the big bang. This BACKGROUND RADIATION was discovered in 1965.
Globular cluster is a ball-shaped aggregation of perhaps 100,000 stars. They are arranged in halos around many galaxies, including our own, where approximately 125 are known. They consist of old stars, up to 14,000 million years old, presumably formed while the galaxies were originally condensing from gas.

H Herschel, Sir John (1792-1871) was an English astronomer who made

compiled famous catalogues of these heavenly bodies, giving them numbers that are still used today. Sir William Herschel made a complete survey of the northern sky and his son Sir John HERSCHEL followed with a survey of the southern sky. Between them they covered the whole of the heavens.

In the early 19th century, the heavens really began to reveal themselves to man. Using bigger and better telescopes, Herschel observed minute movements of the stars and concluded in 1805 that the Solar System is itself in motion towards the constellation Hercules. He also looked at the nebulae discovered by Messier and found that while some of them were indeed glowing clouds of gas, others were distant clusters of stars. Finally, by studying the MILKY WAY – the band of stars that crosses the sky – Herschel found that the Sun is just one of millions of stars in a great group of stars shaped like a grindstone. When we look at the Milky Way, we are looking into this group of stars, which we now call the galaxy. Herschel suspected that some of the nebulae might be other galaxies similar to our own. His son discovered that the Magellanic Clouds consist of thick clusters of stars, and we now know that they are small companion galaxies to our own.

Expanding horizons

Confirmation of Sir William Herschel's speculations had to await the development of very powerful telescopes, which took place at the beginning of the 20th century. From 1912, the American astronomer Harlow SHAPLEY turned the new 1·5-metre and 2·5-metre telescopes at Mount Wilson on the GLOBULAR CLUSTERS, using the CEPHEID VARIABLE method (*see page 23*) developed by Henrietta LEAVITT to find their distances. Shapley found that they lie in a ring around the edge of our galaxy, and thus he was able to find the position of the centre of the galaxy. Herschel believed that we are at the galactic centre but Shapley discovered that we are some 70 per cent of the way out towards the rim of the galaxy. Thus, neither the Sun nor the Earth occupy a central position in the heavens.

Shapley's work marked the beginning of galactic astronomy, which has revealed the Universe as being far, far greater than anyone could have suspected. The foundations of our

present knowledge of the far reaches of the Universe were laid by another American, Edwin HUBBLE. Using the 2·5-metre telescope, Hubble showed in 1924 that a nebula in the constellation of Andromeda is another galaxy like our own. He measured the distance of the ANDROMEDA GALAXY as nearly a million LIGHT YEARS away; in fact, it is more than twice this. Hubble went on to discover many more galaxies and classify them according to their shape. The Andromeda galaxy and our galaxy are SPIRAL GALAXIES, because their stars lie along arms that spiral outwards from a central hub of stars. There are also BARRED SPIRAL

Below: The barred spiral galaxy NGC 1300 in Eridanus. Barred spiral galaxies have a bar-shaped central group of stars from which spiral arms radiate out.

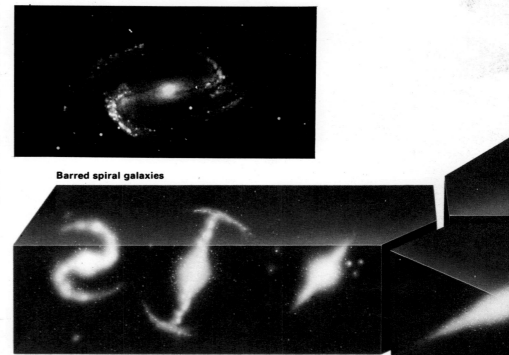

Barred spiral galaxies

SBc   SBb   SBa

E7

Above: The classification of galaxies proposed by Edwin Hubble is used today. Spiral galaxies are indicated by S and barred spirals by SB. Development of spiral arms is shown by adding a, b or c. Elliptical galaxies are denoted by E with a number progressing from 7 to 0 as they become more spherical. Evolution of a galaxy may proceed through the spirals or barred spirals to the elliptical galaxies.

---

the first detailed survey of the southern sky, observing from the Cape of Good Hope from 1834-38. This completed the survey of the entire sky begun by his father William Herschel.
Herschel, Sir William (1738-1822) was an English astronomer, German-born, who on March 13, 1781, discovered the planet Uranus, and later discovered two of its satellites plus two satellites of Saturn. Herschel made a complete survey of the northern skies, discovering many double stars and nebulae. To help with this

work he built a 122-cm reflector, then the largest telescope in the world. Herschel's sky surveys convinced him that our Galaxy was a

Herschel's telescope

lens-shaped system of stars, with our Sun near the centre; this view was accepted until the time of Harlow SHAPLEY.
Hipparchus (2nd century BC) was a Greek astronomer who is considered to have been the greatest of his age. He made an accurate catalogue of 850 stars in which he divided the stars into six brightness groups or magnitudes, the brightest stars being magnitude 1 and the faintest (visible to the naked eye) of magnitude 6. A modified form of this magnitude system is still used today.

Hipparchus discovered that the Earth wobbles slightly in space, an effect known as precession (*see page 43*).
Hoyle, Sir Fred (1915-) is an English astronomer best known for his work on the STEADY STATE THEORY of the Universe, which denies that the Universe began in a BIG BANG. Hoyle has shown how the heavy chemical elements in the Universe are built up from hydrogen and helium by nuclear reactions inside stars, and are distributed into space by SUPERNOVA explosions.
Hubble, Edwin (1889-1953)

was an American astronomer who showed in 1924 that galaxies exist outside our own. He went on to classify galaxies according to whether they were spiral or elliptical in shape. In 1929 he announced that the Universe is expanding, and that the galaxies are rushing away from each other at speeds that increase with their distance; this relationship is known as Hubble's law. A galaxy's distance can be calculated by Hubble's law when its speed of recession is measured from the RED SHIFT of its light. Accord-

Right: The spiral galaxy NGC 5194 in Canes Venatici. Spiral galaxies have a central hub of stars from which 2 or more arms of stars spiral out. This spiral galaxy is notable because a new galaxy is forming at the end of one of the arms and breaking away.

Spiral galaxies

GALAXIES and ELLIPTICAL GALAXIES.

Hubble used the Cepheid variable method to find the distance of the Andromeda galaxy. This method measures the variation in brightness of certain stars known as Cepheid variables. By measuring the apparent brightness as seen from the Earth, their distance can be found. However, this method was of no use where Cepheid variables could not be detected, as in a far distant galaxy. Instead, Hubble used the RED SHIFT in the spectra of light from galaxies first detected by the American astronomer Vesto Slipher in 1912. A galaxy with a red shift in its spectrum is moving away from us. Hubble proposed that the speed of recession, which is proportional to the degree of red shift, is related to the distance of the galaxy. This means that the farther away a galaxy is from us, the faster it is receding from us. This relationship is known as HUBBLE'S LAW, and it shows that the Universe is expanding in all directions. It also shows the Universe to be vast; the farthest known object could be as much as 16,000 million light years away.

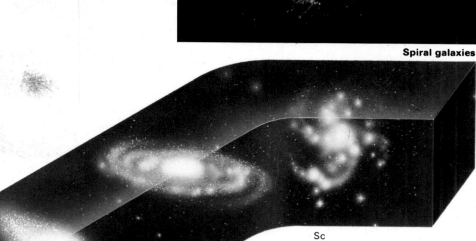

Sc

Sb

Sa

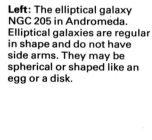

Left: The elliptical galaxy NGC 205 in Andromeda. Elliptical galaxies are regular in shape and do not have side arms. They may be spherical or shaped like an egg or a disk.

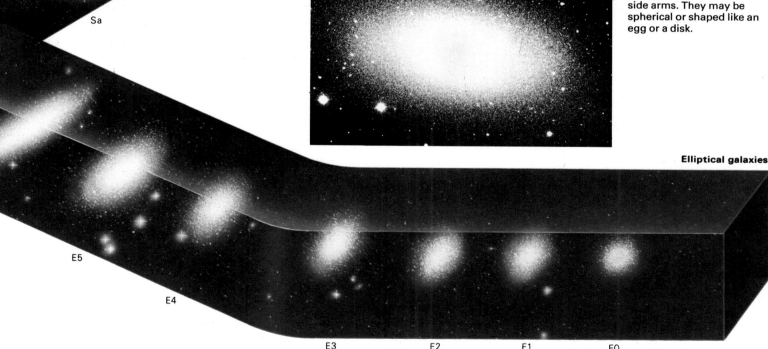

Elliptical galaxies

E6 E5 E4 E3 E2 E1 E0

ing to the latest measurements, galaxies move at 15 km/sec for every million light years of distance.

K Kepler, Johannes (1571-1630) was a German mathematician and astronomer who discovered the three basic laws of planetary motion. First, and most important, he showed in 1609 that planets move around the Sun in elliptical orbits, not with combinations of circles as had always been assumed. He also showed that a planet's speed varies along its orbit,

moving faster when nearest to the Sun and more slowly when farther away. In 1619 he showed that the time a planet takes to complete one

Johannes Kepler

orbit is linked to its average distance from the Sun. For his calculations, Kepler used the observations of Tycho BRAHE.

L Leavitt, Henrietta (1868-1921) was an American astronomer who devised an important technique for measuring distances in astronomy using CEPHEID VARIABLE stars *(see page 30).* She found in 1912 that the average brightness of a Cepheid is directly related to the time it takes to vary, with the brightest Cepheids having the longest periods. By

Henrietta Leavitt (right)

timing the light variations of a Cepheid, astronomers would therefore deduce its true brightness, and thus work out its distance from

other stars and planets.
Lemaître, Georges (1894–1966) was a Belgian astronomer who in 1927 proposed the BIG BANG theory of cosmology, which says that the Universe began in a giant explosion long ago and that the pieces have been flying apart ever since. Lemaître based his theory on the observations Edwin HUBBLE of the expansion of the Universe.
Light year is a measurement of distance astronomy. It is beam of light year. A l

Oscillating theory

Big bang theory

Steady state theory

The origin of the Universe

Hubble's picture of the expanding Universe had a dramatic impact on astronomy. Not only did it give dimension and structure to the Universe for the first time, but it also gave a clue to its formation. This was seized on by the Belgian astronomer Georges LEMAITRE in 1927. He pointed out that if the Universe is expanding now, then it must have been smaller in the past, and at some distant time the galaxies were gathered together in a single mass. This mass exploded, scattering matter into space that formed galaxies as it went. The Universe was created in this BIG BANG and has been expanding ever since. In 1938, the American physicist Hans Bethe worked out that stars produce their energy by converting hydrogen into helium in the process of thermonuclear fusion. Another American, George GAMOW, went on to apply this idea to the big bang, stating that it should have produced helium from hydrogen. Observations later confirmed this prediction.

However, not all astronomers were convinced. Some found it hard to believe that the Universe

Above: The oscillating theory of the Universe holds that the galaxies in the Universe come together until a vast explosion – the big bang – occurs. Then all the matter flies apart, forming new galaxies that move away from each other. This expansion slows and eventually stops, to be followed by a contraction, another explosion, and so on.

Centre: The big bang theory holds that only one part of one cycle has occurred – an explosion about 20,000 million years ago followed by expansion ever since. There is good evidence to support this theory.

Bottom: The steady state theory has proposed that expansion continues indefinitely. New galaxies come spontaneously into being to fill the spaces created by expansion. There is little evidence to support this theory.

could have been created at a definite time in the past. If it is expanding, it could have an infinite future; therefore, why not an infinite past? In 1948, three astronomers put forward an alternative theory of the nature of the Universe. They were the British astronomers Sir Hermann BONDI and Sir Fred HOYLE, and the American astronomer Thomas Gold. Their theory was called the STEADY STATE THEORY.

This theory holds that the Universe had no particular beginning and will never end. The galaxies will continue to move apart for ever, but new galaxies will be created spontaneously to fill the spaces between. Slow creation of matter is taking place continuously, so that the density of the Universe — the number of galaxies in a given space — remains the same. As the galaxies move away from us, they eventually recede beyond the range of our instruments. The Universe could be infinite in extent, meaning that the process has been going on for ever.

Elegant though the steady state theory is, it has not attracted many followers. This is because no evidence has been found to support it,

---

lent to 9,460,000 million km. A parsec is approximately 3·26 light years.

Greater Magellanic Cloud

M **Magellanic Clouds** are two small companion galaxies of our Milky Way, visible as faint fuzzy patches in the Southern Hemisphere of the sky. They are both about 160,000 light years away. The largest contains about 1,000 million stars, 1 per cent of the number in the Milky Way, while the smaller cloud has only about 300 million.

Messier, Charles (1730–1817) was a French astronomer who compiled a list of over 100 star clusters and nebulae. Many of these objects are still referred to by their Messier or M numbers, such as M1, the Crab nebula, and M31, the Andromeda galaxy.

Milky Way is the common name for our galaxy. The Milky Way, seen edge-on from Earth, forms a band of faint light stretching across the sky at night, composed of stars in our galaxy so far away that they are invisible individually without a telescope. There are estimated to be 100,000 million stars in the Milky Way. Our galaxy's diameter is about 100,000 light years, and we are situated over 60% of the way to the rim. The Milky Way galaxy is believed to be a spiral, like the ANDROMEDA GALAXY.

N **Nebula** is a mass of dust and gas in the galaxy. Some nebulae are bright, shining by the light of stars embedded within them; the Orion nebula is one such example. Others are dark, such as the Coal Sack.

Newton, Sir Isaac (1642–1727) was an English scientist whose laws of gravity helped to explain why planets moved around the Sun. Johannes KEPLER had also calculated this by using his laws of planetary motion. Newton also made important contributions to obser-

whereas the big bang theory has undergone the classic test of any theory — a prediction of the theory has been shown to be true. George Gamow, in addition to saying that the big bang would have produced helium, also worked out that some of its heat would remain in the Universe as an all-pervading radiation — like the slight warmth from a dying fire. He predicted the temperature of this BACKGROUND RADIATION, and in 1965 radiation of the correct temperature was detected by radio telescopes. This discovery has remained a powerful piece of evidence for the big bang theory, and most astronomers now believe that the galaxies were created in a big bang about 20,000 million years ago.

However, the big bang theory does not answer all our questions about the origin of the Universe. How did a single mass containing all the matter in the Universe come to exist in the first place? And will the Universe continue to expand for ever?

To answer such questions, another theory of the Universe has been devised. It states that the big bang did take place, but that the Universe will not always continue to expand. The speed of expansion could be slowing, due to the pull of gravity of the galaxies in the Universe. At some point in the future, the expansion could stop and then gravity would begin to pull the galaxies together. The Universe would contract until the galaxies came together in a single mass, and this mass would then explode to create another expansion. This would be followed by another contraction, then another big bang, and so on, in an infinite cycle of oscillations. This theory of the OSCILLATING UNIVERSE also sounds plausible, but there is no hard evidence to support it. Observations have been made that could indicate a slowing down of the expansion of the Universe. If correct, our present cycle of the Universe has another 100,000 million years or so to go.

Quasars

As astronomers debated the nature and origin of the Universe, they found their ingenuity also tested by a series of unusual discoveries. First, in 1940, SEYFERT GALAXIES were found. These galaxies have compact but very bright centres and they emit great energy in the infra-red region of the spectrum. With the development of radio

Sir Isaac Newton

vational astronomy with his research on light and optics. In 1668 he built the first known reflecting telescope.

O Olbers' paradox i famous problem in tronomy put forward in 18 by the German astronom Heinrich Olbers (1758–184C It poses the question: if th Universe is infinite, in every direction you look in space you should see a star o galaxy – so why is the night sky dark? The problem was resolved when it was discovered that the Universe is expanding, so the RED SHIFT weakens the light from distant galaxies.
Oscillating Universe is the theory that the current expansion of the Universe may

Galaxies in Hercules

R Radio galaxy is a galaxy that emits as much as a million times the radio energy of our own galaxy. The radio emission comes from giant clouds on either side of the visible galaxy, which appear to have been ejected from the galaxy's centre in explo sions. The central p source in radio gal believed to be a black hole (se Radio galaxi be a la life of **Rec**

Q Quasar is an intensely brilliant object far off in space believed to be the forming galaxy. Quasars are so small that centre are Quasars appear like stars in even they give off thousands of the largest telescopes, yet they as much energy as a the like our own Milky times may be powered galaxy.They falling into a giant Way gas hole at their centres. by black have the largest RED black shifts. Quasars of any known quasars are several hundred the farthest is be- known; the 16,000 million lieved to be away. light years

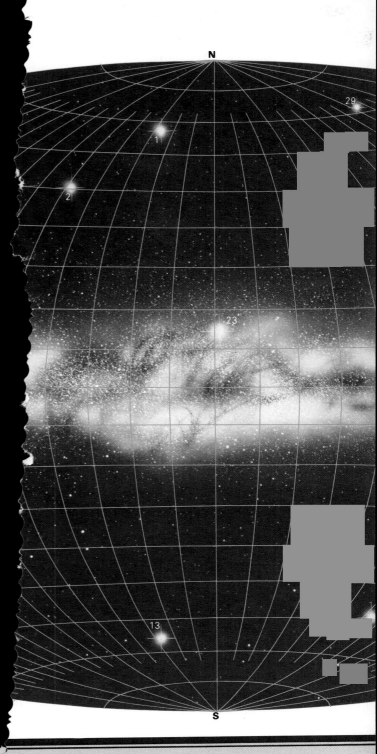

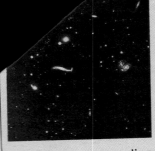

speed of motion. The discovery of the red shift in light from distant galaxies reveals that the Universe is expanding (HUBBLE'S law).

Rosse, Lord (1800–1867) was an Irish astronomer who in 1845 built the largest telescope of its day, a 183-cm reflector, at his home at Birr Castle, Parsonstown. With this telescope Rosse discovered that many of the so-called nebulae in the sky, discovered by William HERSCHEL, were spiral in shape. This was a clue to the fact that they are actually separate galaxies, a fact finally established by Edwin HUBBLE. Rosse gave the Crab nebula its name.

S Schmidt, Maarten (1929–) is an American astronomer who discovered the distances of QUASARS in the Universe. In 1963 he first measured the RED SHIFT of a quasar, 3C 273, finding it to be so great that, according to HUBBLE'S law, it must lie far beyond our own galaxy.

Seyfert galaxy is a type of spiral galaxy with a small, bright nucleus. Seyfert galaxies are believed to be closely related to QUASARS,

ceived from a receding object, caused by the DOPPLER EFFECT. The degree of red shift reveals the object's

...wer ...xies is ...(see page 32). ...es may actually ... stage in the ... QUASARS.

... **shift** is a lengthening of ... wavelength of light re-

whereas the big bang theory has undergone the classic test of any theory — a prediction of the theory has been shown to be true. George Gamow, in addition to saying that the big bang would have produced helium, also worked out that some of its heat would remain in the Universe as an all-pervading radiation — like the slight warmth from a dying fire. He predicted the temperature of this BACKGROUND RADIATION, and in 1965 radiation of the correct temperature was detected by radio telescopes. This discovery has remained a powerful piece of evidence for the big bang theory, and most astronomers now believe that the galaxies were created in a big bang about 20,000 million years ago.

However, the big bang theory does not answer all our questions about the origin of the Universe. How did a single mass containing all the matter in the Universe come to exist in the first place? And will the Universe continue to expand for ever?

To answer such questions, another theory of the Universe has been devised. It states that the big bang did take place, but that the Universe will not always continue to expand. The speed of expansion could be slowing, due to the pull of gravity of the galaxies in the Universe. At some point in the future, the expansion could stop and then gravity would begin to pull the galaxies together. The Universe would contract until the galaxies came together in a single mass, and this mass would then explode to create another expansion. This would be followed by another contraction, then another big bang, and so on, in an infinite cycle of oscillations. This theory of the OSCILLATING UNIVERSE also sounds plausible, but there is no hard evidence to support it. Observations have been made that could indicate a slowing down of the expansion of the Universe. If correct, our present cycle of the Universe has another 100,000 million years or so to go.

Quasars

As astronomers debated the nature and origin of the Universe, they found their ingenuity also tested by a series of unusual discoveries. First, in 1940, SEYFERT GALAXIES were found. These galaxies have compact but very bright centres and they emit great energy in the infra-red region of the spectrum. With the development of radio

Sir Isaac Newton

O **Olbers' paradox** is a famous problem in astronomy put forward in 182? by the German astrono Heinrich Olbers (1758 It poses the questio Universe is infinit direction you you should gala

vational astro
research on
In 1668
known r

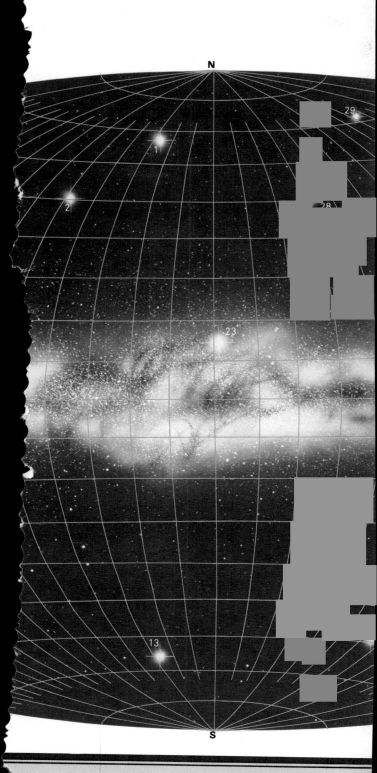

N

29

1

2

28

23

13

S

speed of motion. The discovery of the red shift in light from distant galaxies reveals the Universe is expanding (see HUBBLE's law).

...rd (1800–1867) ...sh astronomer ...uilt the largest ...day, a 183- ...at

established by Edwin HUBBLE. Rosse gave the Crab nebula its name.

S **Schmidt**, Maarten (1929–) is an American astronomer who discovered the distances of QUASARS in the Universe. In 1963 he first measured the RED SHIFT of a quasar, 3C 273, finding it to ...so great that, according ...RBLE's law, it must lie far ...ur own galaxy.

...axy is a type of ...ith a small, ...s. Seyfert ...d to be ...QUASARS,

...eventually slow down and stop, and that the Universe will then begin to contract. According to this theory, the Universe may continue with continuous cycles of expansion and contraction. However, current observational evidence does not support the theory.

P **Ptolemy** (2nd century AD) was a Greek scientist who formulated the standard picture of the Universe used by astronomers until the Renaissance. According to Ptolemy, the Sun, Moon, and planets went around the Earth, their motions being explained by a complicated system of deferents and EPICYCLES. Beyond them was the sphere of the stars which

Ptolemy

revolved around the Earth once a day. This theory was eventually challenged and disproved by the views of COPERNICUS. Ptolemy wrote a major encyclopedia of Greek astronomy, known as the *Almagest.*

Pythagoras (6th century BC) was a Greek scientist who made the first known proposal that the Earth is a sphere. He believed that the Earth lay at the centre of the Universe, and that the other celestial bodies moved around it. *See also* ARISTOTLE; ARISTARCHUS; HIPPARCHUS.

...ó ...mer ...–1840). ...on: if the ...te, in every ...ook in space ...see a star or ...y – so why is the night sky dark? The problem was resolved when it was discovered that the Universe is expanding, so the RED SHIFT weakens the light from distant galaxies.

Oscillating Universe is the theory that the current expansion of the Universe may

...nomy with his ...light and optics. ...he built the first ...eflecting telescope.

Q Quasar is an intensely brilliant object far off in space, believed to be the centre of a forming galaxy. Quasars are so small that they appear like stars in even the largest telescopes, yet they give off thousands of times as much energy as a galaxy like our own Milky Way. They may be powered by gas falling into a giant black hole at their centres. Quasars have the largest RED SHIFTS of any known objects. Several hundred quasars are known; the farthest is believed to be 16,000 million light years away.

R Radio galaxy is a galaxy that emits as much as a million times the radio energy of our own galaxy. The radio emission comes from giant clouds on either side of the visible galaxy, which appear to have been ejected from the galaxy's centre in explosions. The central power source in radio galaxies is believed to be a massive black hole (*see page 32*). Radio galaxies may actually be a later stage in the life of QUASARS.

Red shift is a lengthening of the wavelength of light re-

Galaxies in Hercules

ceived from a receding object, caused by the DOPPLER EFFECT. The degree of red shift reveals the object's

fro___
that t___
ing (HUBL___
Rosse, ___
was an Iris___
who in 1845 b___
telescope of its ___
cm reflector, at his home ___
Birr Castle, Parsonstown. With this telescope Rosse discovered that many of the so-called nebulae in the sky, discovered by William HERS-CHEL, were spiral in shape. This was a clue to the fact that they are actually sepa-*rate* galaxies, a fact finally

c___
be ___
to HUBL___
beyond c___
Seyfert gal___
spiral galaxy ___
bright nucleus ___
galaxies are believe___
closely related to Q___

1	Arcturus	10	Delphinus	20	Betelgeuse
2	Corona Borealis	11	M31	21	Sirius
3	Polaris	12	Pleiades	22	Crux
4	Vega	13	Formalhaut	23	Antares
5	Lyra	14	Achernar	24	Procyon
6	Capella	15	Nebecula Major	25	Castor
7	Cassiopeia	16	Canopus	26	Pollux
8	Deneb	17	Rigel	27	Regulus
9	Altair	18	Aldebaran	28	Spica
		19	M42	29	Denebola

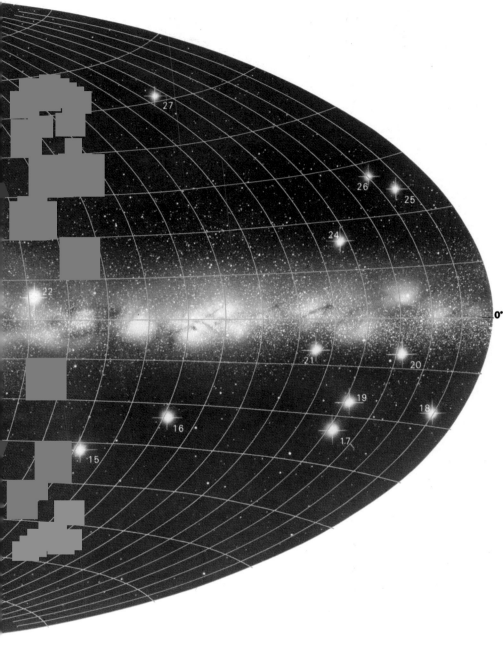

telescopes, astronomers then discovered RADIO GALAXIES, which pump out vast amounts of radio energy. It was hard to explain why these bodies should produce so much energy. But in 1963, a real puzzle presented itself to astronomers.

Three years earlier, the first of several small star-like bodies that are strong radio sources had been found. They were called QUASARS — short for quasi-stellar (star-like) radio sources — and some emitted visible light. In 1963, the American astronomer Maarten SCHMIDT measured the RED SHIFT of a quasar and found that it was very high, placing it about 2,000 million light years away. However, it was several hundred times as bright as a galaxy should be at that distance, and several thousand times smaller! Astronomers checked the red shifts of other quasars and found others even farther away and even smaller. Some are the farthest objects that we can detect and mark the edge of the known Universe, about 16,000 million light years away.

The discovery of quasars offered an explanation of the other bodies. Perhaps both Seyfert and radio galaxies are galaxies that have evolved from quasars and therefore produce large amounts of energy. But what kinds of bodies could quasars be? How could they be so small and yet produce such vast outpourings of energy? Some astronomers doubted that Hubble's Law, which is used to calculate their distances, was correct and that a large red shift does not imply great distance. If this were so, quasars would be much nearer than thought and the problem would be solved. Perhaps a quasar could be a fast-moving body ejected from our own galaxy. Alternatively, gravity affects light and a red shift could be produced by a strong gravitational field. Perhaps such fields lie between quasars and our telescopes.

However, few astronomers are convinced by these theories. A new explanation, though fantastic, seems more likely. Theoretical astronomers have proposed that a star at the end of its life may collapse to such a degree that it becomes a black hole. Its gravity becomes so strong that nothing, not even light rays, can leave it and it appears totally black. Any matter around the black hole will be sucked down into it. As matter disappears in this way, it would give out huge amounts of energy and so a quasar may be a body with a black hole at its centre.

perhaps representing one of the later stages of their evolution into normal galaxies. They are named after the American Astronomer Carl Seyfert who discovered them in 1940.

Shapley, Harlow (1885–1972) was an American astronomer who in 1912 first measured the true size of our galaxy, and showed that the Sun does not lie at the centre. Shapley made his advance from a study of GLOBULAR CLUSTERS of stars, which are distributed in a halo around our galaxy. By measuring their distance from the brightness of the stars they contained, he estimated that the galaxy is about 100,000 light years in diameter, and that the Sun lies about 30,000 light years from the centre.

Spiral galaxy is a galaxy of stars arranged in a spiral shape, like a Catherine wheel. The ANDROMEDA GALAXY is a typical spiral, and so is our own. Spiral galaxies contain between about 1,000 million and one million million stars, and range from 20,000 to over 100,000 light years in diameter. *See also* BARRED SPIRAL GALAXY.

Steady state theory claims that the Universe did not begin with a BIG BANG explosion, but that it never had a specific origin and will never end. In this theory, space has always existed, and new material is slowly created to fill it. As new material spontaneously comes into being, space gets larger and the Universe expands. Therefore an observer would see the Universe looking much the same at any time in the past or future. The steady state theory was proposed in 1948 by Hermann BONDI and Thomas GOLD, and later developed by Fred HOYLE. However, modern observations do not support it.

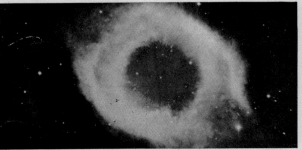

Spiral nebula

Early astronomers relied on their eyesight, but there is now a telescope which can detect candlelight at 24,000 km. Modern radio astronomy has revealed pulsars and quasars and, today, observations can be made from satellites.

Technology in Astronomy

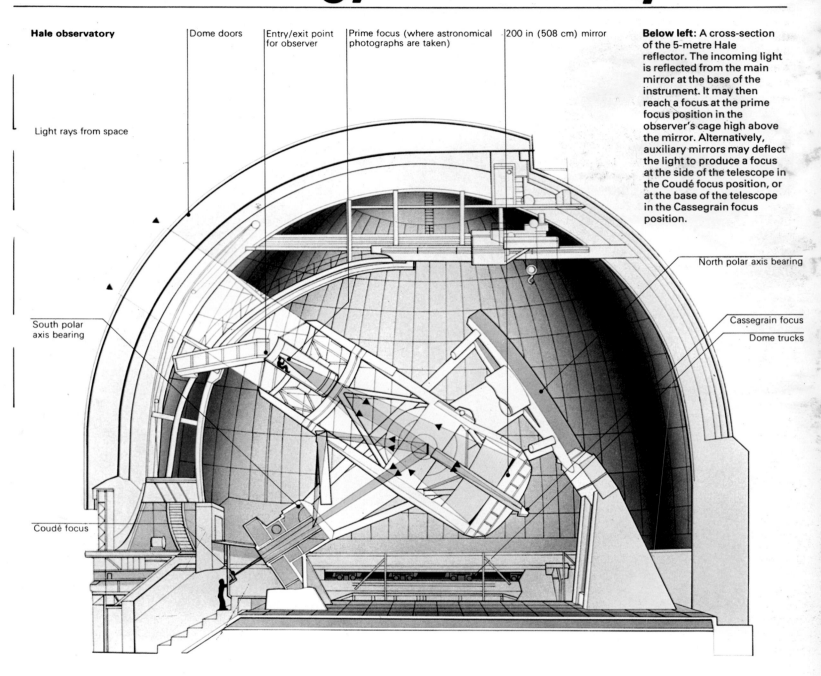

Hale observatory

Dome doors

Entry/exit point for observer

Prime focus (where astronomical photographs are taken)

200 in (508 cm) mirror

Light rays from space

South polar axis bearing

Coudé focus

North polar axis bearing

Cassegrain focus

Dome trucks

Below left: A cross-section of the 5-metre Hale reflector. The incoming light is reflected from the main mirror at the base of the instrument. It may then reach a focus at the prime focus position in the observer's cage high above the mirror. Alternatively, auxiliary mirrors may deflect the light to produce a focus at the side of the telescope in the Coudé focus position, or at the base of the telescope in the Cassegrain focus position.

Reference

A **Altazimuth mounting** is a simple type of mounting that allows a telescope to swivel freely up and down (in *altitude*) and from side to side (in *azimuth*). This type of mounting is widely used for binoculars and small telescopes, but large instruments use an EQUATORIAL MOUNTING.

Anglo-Australian Observatory is an astronomical observatory at Siding Spring, New South Wales, containing the 3.9-metre Anglo-Australian Telescope. At the same site is the 122-cm UK SCHMIDT TELESCOPE, the largest of its kind in the Southern Hemisphere.

Aperture synthesis is a technique used by radio astronomers in which observations by a number of small radio telescopes are combined to produce the view of the sky that would be seen by a single radio dish of much larger aperture.

Arecibo Radio Observatory is the site in Puerto Rico of the largest single radio astronomy dish in the world, 305 metres in diameter. The Arecibo radio telescope lies in a natural hollow between surrounding mountains, from which it is suspended.

Siding Spring Observatory

The dish cannot be steered, but its beam sweeps across the sky as the Earth rotates.

C **Cassegrain telescope,** invented in 1672 by a French physicist, N. Cassegrain, is a type of reflector in which the eyepiece is placed in a hole at the centre of the main mirror. This mirror collects light and focuses it on to a smaller, convex mirror in front, which bounces it back into the eyepiece.

Cerro-Tololo Inter-American Observatory is an observatory on 2,160-metre high Cerro Tololo mountain in Chile containing the largest optical telescope in the Southern Hemisphere. This is a twin to KITT PEAK, Cerro Tololo's sister observatory.

Clark, Alvan (1804–87) was the leading American telescope optician of his day. In 1862 he built a 47-cm refractor with which his son Alvan Graham Clark discovered the white dwarf companion to the star Sirius, and in 1871 completed the 66-cm refractor of the US Naval Observatory, Washington, with which Asaph Hall discovered the two moons of Mars.

The year 1609 is probably the most important date in the history of astronomy. Johannes Kepler published the first of his laws of planetary motion, resolving centuries of argument on the nature of the Universe. And Galileo looked at the sky through a TELESCOPE for the first time. Both events mark the beginning of modern astronomy.

The invention of the telescope took a long time to happen. The necessary lenses had been around for several centuries, and Roger Bacon in the 13th century proposed using combinations of lenses for telescopes. In the end, the telescope was discovered by accident. In 1608, an assistant to a Dutch spectacle maker happened to place one lens in front of another and noticed that they made distant objects look nearer. The spectacle maker, Hans Lippershey, constructed the first instrument for the Dutch government in 1608, and Galileo heard about it shortly after. He immediately set about constructing his own telescope, and made one that could magnify 32 times. With it, he observed the phases of Venus, four of the moons of Jupiter, the rings of Saturn (though he did not recognize them as such) and sunspots.

Before the telescope, astronomers could only record the positions of the stars and planets visible to the naked eye and speculate on their nature. Now they could take a close look at these and discover many more too faint to be seen by the naked eye. The telescope had opened up the Universe.

How telescopes work

A telescope gathers light rays coming from a distant object and forms them into an image of the object. In 'refracting' telescopes, the image is formed by a lens and in 'reflecting' telescopes by a curved mirror. This image is then viewed through a small eyepiece lens. The magnification of the object depends mainly on the power of the eyepiece lens. The telescopes used by astronomers are very large. This is not just to obtain images that are highly magnified, though the length does help, but to gather as much light as possible. The reason we can see stars through a telescope that are too faint to see with the naked eye is that the telescope is wider than the eye and gathers more light. The diameter of the image-forming lens or mirror is therefore always given when describing an astronomical telescope. The

Right: The refracting telescope has lenses to focus the incoming light rays and produce an image. This image is then viewed by an eye-piece lens and magnified.

Right: Galileo's telescope. Galileo did not invent the telescope, but built the first instruments to be used in astronomy in 1609. They were refracting telescopes.

Far right: The first reflecting telescope, built by Isaac Newton in 1668. James Gregory designed such an instrument earlier but never built it.

Right: The reflecting telescope has a mirror or system of mirrors to bring the incoming light to a focus and form an image. This image is then viewed through an eyepiece lens.

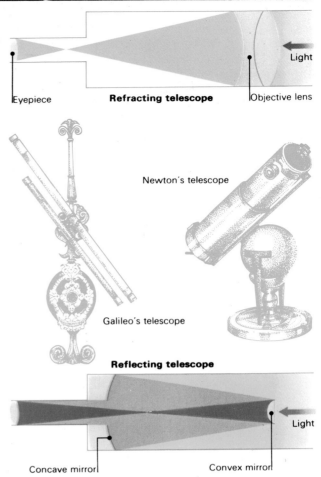

Refracting telescope — Eyepiece — Objective lens — Light

Newton's telescope

Galileo's telescope

Reflecting telescope — Concave mirror — Convex mirror — Light

most powerful telescope in the world, the six-metre reflecting telescope at ZELENCHUKSKAYA OBSERVATORY, can detect the light of a candle 24,000 kilometres away.

The largest refracting telescope is the 102-centimetre refractor at Yerkes Observatory in the United States. It was built in 1897. The lenses used in refracting telescopes need to be made very carefully to avoid distortion of the image, or 'aberration', and large lenses absorb light as it passes through them and are very heavy, making construction difficult. The large and powerful telescopes built in this century are therefore all reflectors. Ensuring that the mirror of an astronomical telescope is free from aberration and absorbs very little light, takes years of polishing. Although it is heavy, the mirror is placed at the base of the telescope and so is easy to support.

D **Dollond,** John (1706–61) was an English optician who in 1757 discovered how to make telescope lenses that did not produce the red and blue colour fringes known as chromatic aberration that marred the images of ordinary refracting telescopes. Dollond's so-called achromatic lenses were made by placing together lenses made of 2 different types of glass. This principle was used by later lensmakers.

E **Effelsberg Radio Observatory,** near Bonn, West Germany, is the site of the world's largest fully steerable radio telescope, the 100-metre dish of the Max-Planck-Institut für Radioastronomie, which began operating in 1971.
Equatorial mounting is a telescope mounting in which one axis is aligned parallel with the axis of the Earth. Once the telescope has been pointed at an object in space, this can be kept in view without further adjustment, by slowly turning the telescope's axis to counteract the effect of the Earth's rotation.

G **Gregory,** James (1638–75) was a Scottish mathematician and inventor who in 1663 published the first known design for a reflecting telescope. It used a concave primary mirror and a concave secondary. It has now been superseded by the CASSEGRAIN design.

H **Hale,** George Ellery (1868–1938) was an American astronomer who set up the world's largest refracting telescope, at Yerkes Observatory, and the 2.5-metre and 5-metre reflectors of Mount Wilson and Mount Palomar (now HALE OBSERVATORIES). Hale was particularly interested in the Sun, finding that sunspots were cooler areas on the

George E. Hale

Sun's surface associated with strong magnetic fields.
Hale Observatories has been the name since 1970 of the Mount Wilson and Palomar Observatories in California, founded by George Ellery HALE. On Mount Wilson are the 1.5-metre and 2.5-metre reflectors, opened in 1908 and 1917 respectively. Mount Palomar houses a 122-cm Schmidt photographic telescope and the famous 5-metre reflector, opened in 1948, the largest until the opening of ZELENCHUKSKAYA OBSERVATORY.

The mirror forms an image at the 'focus' of the telescope, and the astronomer must be situated at this point. In the prime focus telescope, the focus is above the mirror and the observer sits in a cage inside the telescope. Other kinds of reflectors have additional mirrors that deflect the light rays from the main mirror to move the focus. In the CASSEGRAIN TELESCOPE, the additional mirror reflects the light back through a hole in the main mirror so that the astronomer works at the base of the instrument, as with a refractor. In Newtonian and Coudé telescopes, the focus is to one side of the telescope. Many reflectors are built so that the focus can be changed to different positions.

Most astronomical telescopes have a narrow field of view; increasing this field requires lenses or mirrors that introduce some distortion of the image. The SCHMIDT TELESCOPE overcomes this problem by using a special lens to correct distortion produced by a mirror.

Observatories

Most observatories are situated at the summits of high mountains, far away from any interference from artificial light and if possible above the thicker, low clouds. As light rays pass through the atmosphere, they may be bent by different layers of air; this makes the stars appear to twinkle. To make the best use of a telescope, it should be as far above sea level as possible.

Each telescope in an observatory is situated beneath a dome that opens as night falls. It is mounted so that the astronomer can direct it at any point in the sky. In addition, it is driven so that it slowly turns to track any particular body through the sky as the Earth rotates. Either ALTAZIMUTH MOUNTING or EQUATORIAL MOUNTING is used to pivot the telescope as it turns.

The modern astronomer does not spend much time looking through telescopes. The human eye is a very limited receiver of information. Photographic film is far more sensitive to light, and faint bodies are detected by taking long-exposure photographs of the skies. The eye is also insensitive to radiations such as ultra-violet rays and infra-red rays that can be recorded by photography to give us valuable information. In addition, photographs taken in particular colours bring out hidden features.

Astronomers also make use of various instruments to extend their knowledge of the Universe. One of the most valuable is the spectroscope, which produces a spectrum of light coming from stars and planets. This is done by spreading the light out into its component colours, as a prism

Below left: Zelenchukskaya Observatory, situated at a height of 2,070 metres in the Caucusus mountains of Russia, houses the world's largest optical telescope. It is a 6-metre reflector which began work in 1976.
Below: This radio telescope in the Crimea in Russia has a set of small dishes that can be steered to pick up signals from any point in the heavens. It is used for tracking spacecraft. Similar but larger dishes are used for radio astronomy.

Kitt Peak Observatory

Sir Bernard Lovell

does with sunlight. From the pattern of its spectrum, astronomers can tell which elements a star is composed of, how fast it is moving or rotating, and how far away it is. Other instruments measure the light coming from a star, telling us how hot it is and how bright it is.

Invisible astronomy

Light is not the only radiation to reach us from space. Although the atmosphere blocks many kinds of radiation from reaching the ground, it does have two 'windows'. One is the optical window, allowing light rays to penetrate to the ground, and the other is the radio window, for radio waves can also penetrate the atmosphere. This was not realized until some time after the development of radio, because the radio waves that reach us from space make only a faint hissing noise in short-wave radio receivers. This noise was first detected by an American radio engineer named Karl JANSKY, who found that it interfered with long-distance telephone calls sent by radio. He traced the source of interference to the galaxy but did nothing more to follow up this discovery, even though it was big news when he published it in 1932.

The development of radio astronomy was vital to the future of astronomy. Light cannot penetrate the dust clouds that obscure the centre of the galaxy, but radio waves can. A radio picture would show the structure of the galaxy, and new bodies would reveal themselves by their radio signals.

The first steps into this exciting new field of discovery were taken not by professional astronomers, but by an amateur radio enthusiast named Grote REBER. He built a special receiver in his back garden in 1937, and with it he made the first, probably incomplete, radio map of the Universe. World War II delayed the development of the new science, but in the 1950s the spiral structure of the Galaxy was shown clearly and radio galaxies – intense sources of radio waves – were discovered. Much of this pioneering work was done in Holland and in Britain, where the first large radio telescope was built at JODRELL BANK by Sir Bernard LOVELL.

Radio telescopes

There are two main kinds of RADIO TELESCOPES, and neither resemble an optical telescope. Both

Above: The dish of the radio telescope at Arecibo, Puerto Rico, is built into a valley. It is 305 metres across and is the largest dish in the world. Although it cannot be steered, it sweeps across the sky as the Earth rotates.

Below: Herstmonceux Castle in Sussex, England, is the new site of the Royal Greenwich Observatory.

cal observatory of the University of Texas, opened in 1939, situated on Mount Locke. Its main telescope is a 2.7 metre reflector installed in 1968.
Mount Wilson and Palomar observatories, see HALE OBSERVATORIES.
Mullard Radio Observatory, situated near Cambridge, England, is operated by the University of Cambridge. Its main telescope consists of eight radio dishes each 12.8 metres in diameter arranged in a line 5 km long, working on the APERTURE SYNTHESIS principle

invented by the observatory's director Martin RYLE. PULSARS were discovered by the observatory in 1967.

O **Object glass** is the lens that collects light at the front of a refractor. The object glass focuses light so it can be magnified by the eyepiece. Object glasses usually consist of at least two lenses made from different types of glass, to prevent the colour fringes (or chromatic aberration).

P **Parkes Observatory,** in New South Wales, is the

national radio astronomy observatory of Australia.
Pic du Midi Observatory is an optical observatory at an altitude of 2,862 metres in the French Pyrenees. Be-

Pic du Midi Observatory

cause of its extreme altitude it has furnished some of the clearest views of the Moon and planets.

R **Radar astronomy** is the technique of bouncing radio waves off a surface and detecting their echo. Radar can be used to measure distances to stars and planets in the solar system very accurately, and also to assess the nature of their surfaces. Radar astronomy revealed the true rotation periods of Mercury and Venus, as well as features on the surface of Venus.

Radio telescope is a device for collecting long-wave radiation from space. Most radio telescopes are in the form of a dish, which col-

Radio telescope dish

Reber's receiver and the Jodrell Bank instrument are dish radio telescopes: a metal dish reflects incoming radio waves to a receiver placed above the dish. In fact, this kind of telescope works in exactly the same way as an optical reflecting telescope, the dish acting like a mirror to the incoming rays and bringing them to a focus at the receiver. The other kind of radio telescope consists of large arrays of separate aerials and receivers, instead of a single dish. It is rather like a dish taken to pieces and spread out over the ground.

Radio waves have much longer wavelengths than light rays, and this means that a radio telescope has to be very large to produce a detailed picture. The radio telescope at EF-FELSBERG RADIO OBSERVATORY in West Germany has the largest steerable dish, which is 100 metres in diameter. Being steerable, the dish

Right: A radio telescope works in much the same way as a reflecting telescope. The incoming radio waves are reflected by the dish and come to a focus at the antenna. There they produce electric signals that go to detecting instruments in the observatory.

Below: Effelsberg Radio Observatory in West Germany is the site of the largest steerable radio telescope dish in the world. It is 100 metres in diameter. The telescope began work in 1971.

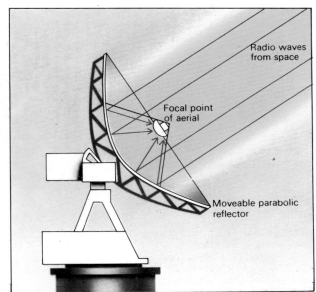

can be pointed anywhere in the Universe. The largest dish of all is at the ARECIBO RADIO OBSERVATORY in Puerto Rico. It is 305 metres across and, being built into a valley, cannot be moved. However, it sweeps the sky as the Earth rotates. Radio astronomers can effectively obtain larger instruments by linking radio telescopes together to form an INTERFEROMETER. This technique is called APERTURE SYNTHESIS and has been particularly successful at the MULLARD RADIO OBSERVATORY in Britain. A huge instrument working on this principle will begin to operate in the United States in 1981. It is called the VERY LARGE ARRAY, and will do the work of a single dish 27 kilometres across.

As the incoming radio signals are brought to a focus at the receiver, they produce weak electric signals in the receiver. These signals are amplified and then recorded on tape or as traces on charts. The radio signals come from the Sun, the planet Jupiter, stars, nebulae and galaxies, revealing more about their nature. Radio astronomy has also revealed new kinds of bodies such as quasars and pulsars. In addition, radio waves come from space. A constant signal distributed uniformly over the Universe that was discovered in 1965 supports the theory that the Universe began with a big bang. Other signals come from molecules of compounds drifting in space. These include water and ammonia, substances involved in the formation of life.

lects radio waves and focuses them onto a receiver.
Reber, Grote (1911-) is an American pioneer of radio astronomy. As an amateur radio enthusiast, he was the first person to follow up Karl JANSKY's discovery of radio noise from the galaxy. Using a 9.4 metre diameter dish in his back garden, Reber plotted the radio brightness of the Milky Way in the 1940s, and detected individual sources, such as the Crab nebula.
Royal Greenwich Observatory was founded in 1675 in London, England, to com-

pile tables of stars and the Moon's motion to assist navigators. Since then its activities have expanded to take in all aspects of modern astronomy. Until 1972, its directors bore the title Astronomer Royal. In 1884 Greenwich was chosen as the world's zero line of longitude (the prime meridian) and Greenwich time became the world's standard. In 1958 the observatory moved to new premises at Herstmonceux Castle in Sussex.
Ryle, Sir Martin (1918-) is a British radio astronomer, director of the MULLARD RADIO

OBSERVATORY, who in 1974 shared the Nobel prize in physics with colleague Antony Hewish for his invention of the APERTURE SYNTHESIS design of radio telescope. Ryle in 1972 became Astronomer Royal.

S Schmidt telescope is a telescope for taking wide-angle photographs of the sky. The telescope has a combined lens and mirror system: light passes through a specially shaped corrector lens at the front onto a mirror, which reflects the light back to a photo-

graphic plate. This combination gives the Schmidt telescope a wide field of view, free from distortion. The design was invented in 1930 by the Estonian optician Ber-

nhard Schmidt (1879-1935)
Seeing is a term given to the steadiness of the atmosphere, which affects the quality of an image as seen through a telescope. 'Good

Greenwich Observatory in Flamsteed's time

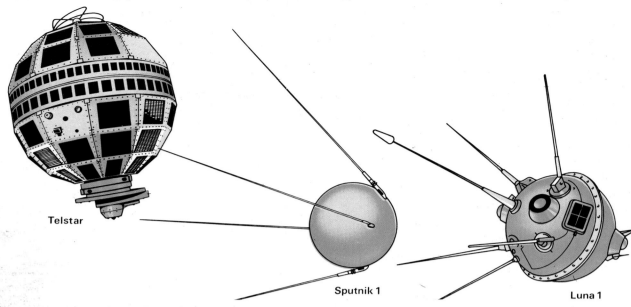

Right: Several pioneering satellites and space probes. Telstar was placed in orbit by the United States in 1962 and relayed the first television pictures across the Atlantic Ocean. Sputnik 1, launched by Russia in 1957, was the first satellite of all. It carried out scientific research. Luna 1 was the first spacecraft to leave the Earth's field of gravity. It was launched by Russia in 1959 and passed near the Moon.

Below: The galaxy M82 in Ursa Major is a strong source of radio waves. Photographs reveal what appears to be an enormous explosion taking place at the centre of the galaxy with jets of hydrogen streaming out in all directions.

Telstar

Sputnik 1

Luna 1

Another use of radio telescopes is to send signals out into space and to listen for any echo or reply. RADAR ASTRONOMY bounces radar waves off the planets to find out more about them, and radio messages have been sent out to the stars in the hope that some day a reply will come back. We are likely to wait a long time. It has been estimated that the possible sites of intelligent life in the galaxy lie about 300 light years apart.

Astronomy in space

Light rays and radio waves are only two of the radiations that come from space. However, all the others, which include X-rays and most ultraviolet and infra-red rays, are prevented from reaching the ground by the atmosphere. Astronomers have been able to make observations of space with these radiations by sending instruments above the atmosphere aboard rockets and satellites. This field of astronomy is still in its infancy, but it has already yielded important results. One of the main tasks of the Skylab astronauts, who orbited the Earth in 1973 and 1974, was to study the Sun in the X-ray and ultraviolet region. They obtained more information about the Sun than astronomers had ever been able to get on Earth. The discovery of X-rays coming from the constellation Cygnus (the Swan), although there is no visible star at this point, could be the first evidence that black holes exist out in space.

seeing' is when the atmosphere is at its steadiest. When there are extensive atmospheric currents, which cause turbulence, the seeing is said to be poor, and the image in the telescope seems to "boil", or dance around.

T **Telescope** is a device for collecting and magnifying light from distant objects. Since telescope lenses or mirrors have a wider opening than the human eye, they can collect more light, and thus show objects too faint to be seen

by the eye alone, and also can obtain finer detail. Refracting telescopes were the first to be made. Their invention is attributed to the Dutch optician Hans Lippershey (1587-1619) in 1608. Reflectors, which use mirrors, were first suggested by James GREGORY, but the first known example was made by Isaac Newton. All large telescopes are now reflectors, because they are cheaper and easier to make than refractors.

U **US Naval Observatory** is a US govern-

ment observatory, the American equivalent of the ROYAL GREENWICH OBSERVATORY, with its headquarters at

Peking Observatory

Washington, D.C., where it operates the 66-cm refractor with which Asaph Hall discovered the moons of Mars.

V **Very Large Array** is the name of the world's largest radio astronomy instrument, under construction in the United States near Socorro, New Mexico. When completed in 1981 it will consist of 27 movable antennae using the technique of APERTURE SYNTHESIS.

Z **Zelenchukskaya Observatory,** at an altitude of 2,070 metres in the Caucasus mountains of Russia, is the site of the world's largest optical telescope, the 6 metre reflector opened in 1976.

The Sun is one of countless millions of stars and is an ordinary, middle-aged star. We can take comfort in this fact, considering what we now know about some of the strange stars at other stages of development in the Universe.

The Stars

Some 6,000 stars are visible to the naked eye, and about a third of this number may be seen from any one place on Earth. Which particular stars will be visible in the night sky at a particular time, depends on the hour of night, the time of year and the observer's position on the surface of the Earth.

As the Earth rotates, the stars appear to move across the sky. In the Northern Hemisphere, they seem to circle around the Pole Star, which is almost directly above the North Pole and therefore does not appear to move. At the Equator, the stars seem to move overhead in straight parallel lines. In the Southern Hemisphere, they apparently circle around a point above the South Pole. Unlike the northern skies, there is no southern star to mark the centre of the circles described by the stars.

During the year, as the Earth goes round the Sun, the stars that are visible in the night sky from a particular place change from one season

to another. Some stars that can be seen in the night sky in winter cannot be seen in summer because they are obscured by daylight, and vice-versa. The observer's location also determines the range of stars that are visible throughout the year. At the Equator all the stars come into view. Elsewhere, the range is more and more limited the farther north or south the observer stands. Stars that remain in view all the year round are known as circumpolar stars.

Constellations

The stars are very far away from us, and although the Earth's motion around the Sun causes PARALLAX, it is so slight that to the naked eye, the position of the stars in relation to one another apparently does not alter. The patterns they form in the sky are static though the planets slowly drift through this backdrop of 'fixed' stars.

The brighter stars have special names, given to them by Arab astronomers many centuries ago. Others are known by the particular CONSTELLATIONS to which they belong. A constellation is a particular pattern of stars in the sky. Usually, the stars of a constellation only appear to be close together; generally, they are far apart in space. The constellations were also named long ago, and have the names of animals, such as the Swan and Lion; heroes of legends, such as Hercules and Perseus; common objects, such as the Scales and the Lyre; and people, such as the Archer and the Sculptor. Often, the constellations are referred to by their Latin names, and their resemblance to the beings or things that they represent is usually difficult to see. The Plough (which is part of the constellation of the Great Bear) does look like a plough and the Southern Cross is cross-shaped, but the Swan is also shaped like a large cross and Cassiopeia, (the name of a mythical queen), looks like a big 'W' in the sky. However, the various patterns have

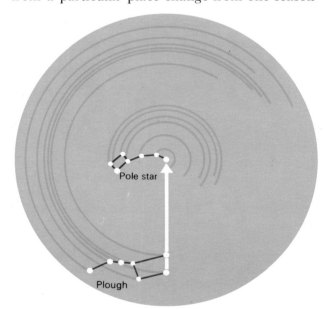

Pole star

Plough

Left: Polaris, or the Pole Star, is situated almost exactly above the North Pole. It therefore does not appear to change position in the sky as the Earth rotates, whereas all the other stars in the northern skies appear to revolve around it. The Pole Star can easily be found by extending a line joining the first 2 stars of the Plough or Ursa Major (the Great Bear) as shown. In addition, the Pole Star forms the last star in the similar constellation called Ursa Minor (the Little Bear).

Reference

A **Absolute magnitude** is a measure of a star's actual light output. It is the brightness of a star as it appears to us on Earth at a distance of 10 parsecs (32.6 light years). A star's absolute magnitude depends on its size and temperature.
Aldebaran is a RED GIANT star, the brightest in the constellation Taurus. Aldebaran is 65 light years away and has a diameter 36 times that of the Sun.

Algol, also called Beta Persei, is an eclipsing binary star 82 light years away in the constellation Perseus. Algol dims slightly every

Signs of the zodiac

2.87 days as it is eclipsed by a fainter companion star. Algol's brightness variations were discovered and explained in 1782 by the English amateur astronomer John Goodricke (1764–86).
Alpha Centauri is the closest star to the Sun, 4.3 light years away. Actually, it consists of 3 stars linked by gravity, although to the naked eye they appear as 1. Alpha Centauri is the brightest star in Centaurus, and the third brightest in the sky.
Antares is a red giant star, nearly 300 times the Sun's

diameter, lying in the constellation of Scorpius. One of the largest stars known, it gives out as much light as 5,000 suns. Antares is 430 light years away.

Aquarius

Apparent magnitude is the brightness of a star or any other celestial object as it appears to us in the sky. A star's apparent magnitude depends on its distance from us – the nearer a given star is, the brighter it appears. The difference between a star's apparent magnitude and its ABSOLUTE MAGNITUDE reveals its distance.
Aquarius, the Water Carrier, is a constellation of the zodiac, lying in the equatorial region of the sky. The Sun passes through the constellation from mid-February to mid-April.

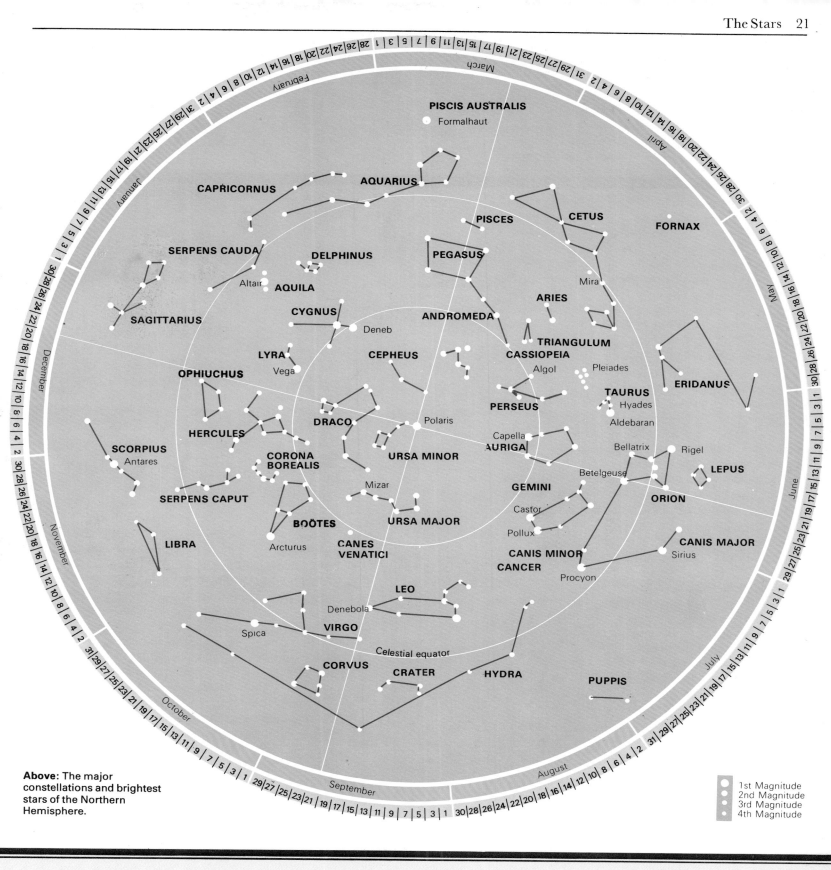

PISCIS AUSTRALIS
Formalhaut

CAPRICORNUS
AQUARIUS
PISCES
CETUS
FORNAX
SERPENS CAUDA
DELPHINUS
PEGASUS
ARIES
Mira
Altair
AQUILA
CYGNUS
ANDROMEDA
TRIANGULUM
Deneb
SAGITTARIUS
CASSIOPEIA
LYRA
CEPHEUS
Algol
Pleiades
Vega
OPHIUCHUS
ERIDANUS
TAURUS
PERSEUS
Hyades
DRACO
Polaris
Aldebaran
HERCULES
Capella
AURIGA
Bellatrix
Rigel
SCORPIUS
CORONA
BOREALIS
URSA MINOR
Betelgeuse
LEPUS
Antares
Mizar
GEMINI
ORION
SERPENS CAPUT
Castor
CANIS MAJOR
BOÖTES
URSA MAJOR
Pollux
LIBRA
Arcturus
CANES
VENATICI
CANIS MINOR
Sirius
CANCER
Procyon
LEO
Denebola
CANIS MAJOR
Spica
VIRGO
Celestial equator
CORVUS
CRATER
HYDRA
PUPPIS

Above: The major constellations and brightest stars of the Northern Hemisphere.

1st Magnitude
2nd Magnitude
3rd Magnitude
4th Magnitude

Arcturus is a red giant star, 27 times the Sun's diameter, lying in the Northern Hemisphere constellation Boötes, the Herdsman.

Open cluster in Cancer

Aries, the Ram, is a faint constellation of the zodiac, lying in the Northern Hemisphere of the sky. The Sun is within its boundaries from late April to mid-May each year.

B **Barnard's star** is the second closest star to the Sun, 6 light years away. It lies in the Northern Hemisphere constellation Ophiuchus, the Serpent Bearer, and is so faint – it is a WHITE DWARF – that it is invisible without a telescope. Barnard's star is named after the American astronomer

Edward Emerson Barnard (1857–1923) who in 1916 discovered that it has the greatest PROPER MOTION of any star. It is possible that Barnard's star is accompanied by 2 planets, roughly the size of Jupiter and Saturn.

Betelgeuse is a red super giant star marking the right shoulder of the constellation Orion. Betelgeuse is so large that it is unstable, changing irregularly in diameter between 300 to 400 times the size of the Sun and varying slightly in brightness as it does so. Betelgeuse lies 650 light years away.

Binary stars (or double stars) are a pair of stars linked by gravity. Most binaries appear as single stars to the naked eye, but can be seen separately through telescopes, but some are so close that their existence can only be deduced from analysis of their combined light. In some binaries the stars periodically eclipse each other.

Black holes are areas of space surrounding a collapsed star in which gravity is so strong that nothing can escape, not even light, although things can be sucked

in. If the big bang explosion marked the origin of the Universe, much smaller black holes might have been formed in the conditions of

Aries

definite and distinctive shapes which can be recognized even though their names are not immediately obvious.

The positions of the constellations are shown on star maps. The maps are usually circular, one representing the skies of the Northern Hemisphere and the other the Southern Hemisphere. They show all the constellations that lie overhead, and must be imagined as stretching from one horizon, over one's head to the opposite horizon. The two maps overlap, because the constellations at the edges can be seen from either hemisphere depending on the time of year.

A star map joins up the stars into their constellations and names the brighter ones. In fact, all the brighter stars are named after their own constellation, using letters of the Greek alphabet to identify them in order of their brightness. Thus BETELGEUSE and Rigel are known as Alpha Orionis and Beta Orionis, being the two brightest stars in the constellation of ORION (although since the stars were first named in this way, the order of brightness has sometimes been found to be incorrect: Rigel is in fact brighter than Betelgeuse, for example).

Across the star maps a wavy band is sometimes shown. This represents the Milky Way, the belt of stars that lies across the night sky and which consists of the stars lying in the plane of our Galaxy. There may also be a grid of lines with figures to denote the positions of the stars; the grid gives the DECLINATION and RIGHT ASCENSION of any star — the celestial equivalent of latitude and longitude.

Sights of the northern skies

The best-known constellation of the northern skies is probably the Plough. Its seven bright stars stand out at all times, three of them forming the curved handle of the plough and the other four stars the ploughshare. The central star in the handle, Mizar, is in fact a double star; it has a companion known as Alcor. This pair is one of the few that can be seen with the naked eye. The Plough is also important as a signpost. The two stars at the end of the ploughshare are known as the 'pointers', because they point towards to the Pole Star or POLARIS. ARCTURUS, one of the brightest stars in the sky, is found by imagining a continuation of the curve of the handle through the sky. Near Arcturus is the Northern Crown, a

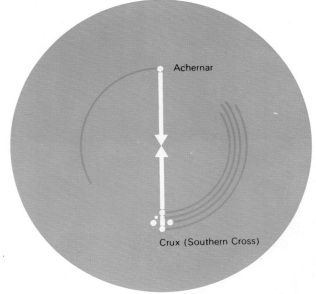

Right: The southern celestial pole lies midway between the constellation Crux (the Southern Cross) and Archernar, the brightest star in the constellation Eridanus.

distinctive 'C'-shaped curve of stars.

The Pole Star itself makes up the end of the handle of a miniature plough that marks the constellation of the Little Bear or URSA MINOR (the Plough being part of the Great Bear or URSA MAJOR). On the opposite side of the Pole Star to the Plough is the 'W' shape of CASSIOPEIA. Carrying on in the same direction, we come to the mythical daughter of Cassiopeia, Andromeda. This constellation is unremarkable, but it contains a fuzzy patch that through a telescope reveals itself to be another galaxy. Next to Andromeda lies Pegasus (the winged horse of mythology), its main features being a great square marked out by four stars.

Returning to Cassiopeia, we find ourselves in the Milky Way and, to the right of the 'W' of Cassiopeia, we come to the Swan or CYGNUS. This constellation lies like a giant cross along the Milky Way and represents a swan flying along the belt of stars. On the other side of Cassiopeia in the Milky Way lies PERSEUS with the star ALGOL, which varies regularly in brightness, and beyond is Auriga, the Charioteer, with the very bright star CAPELLA. Beside Auriga and alongside the Milky Way is TAURUS, the Bull, which contains the red star ALDEBARAN. Nearby is the famous star cluster PLEIADES, also known as the Seven Sisters because most people can see seven stars in the cluster (although there are in fact many more).

---

high density and pressure which followed. There is at present no conclusive evidence for the existence of a black hole.

C **Cancer,** the Crab, is a constellation of the zodiac, lying in the Northern Hemisphere of the sky. The Sun passes through the constellation from late July to mid-August. Cancer is the faintest of the zodiacal constellations, with no bright stars. Its most interesting feature is the star cluster Praesepe, known as the Beehive, 520 light years away.

Canopus is the second brightest star in the sky. It lies 110 light years away in the constellation Carina, the Keel, in the Southern Hemisphere. Canopus is a yellow giant, 25 times the diameter of the Sun. Space probes use it as a navigation star.
Capella is the brightest star in the Northern Hemisphere constellation Auriga, the Charioteer. It is actually a binary star, consisting of 2 components orbiting each other every 104 days. Capella is 45 light years away.
Capricornus, the Sea Goat, is a constellation of the

zodiac, lying in the Southern Hemisphere of the sky. The Sun passes through the constellation from late January to mid-February.

Capricorn

Cassiopeia is a prominent 'W' shaped constellation near the north pole of the sky, named after a Greek mythological queen. In this

constellation a bright SUPERNOVA flared up in 1572, observed by Tycho Brahe.
Castor is the second brightest star in Gemini, a

Galaxy in Centaurus

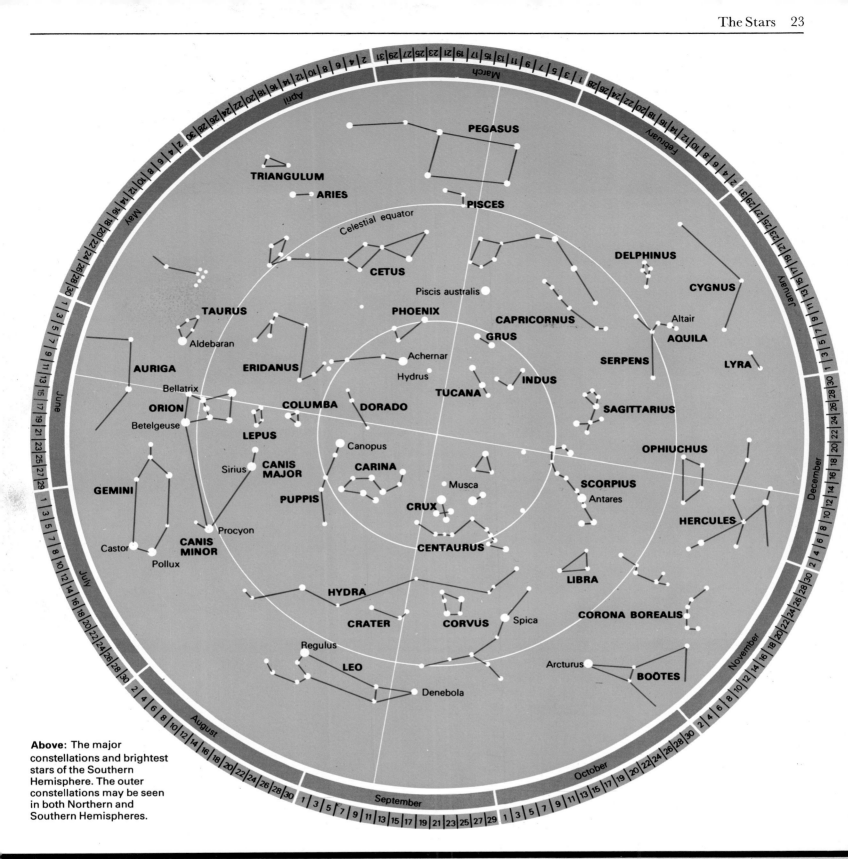

Above: The major constellations and brightest stars of the Southern Hemisphere. The outer constellations may be seen in both Northern and Southern Hemispheres.

Northern Hemisphere constellation. Castor actually consists of 6 stars linked by gravity, although they appear as 1 star to the naked eye.

Centaurus, the Centaur, is a prominent constellation in the Southern Hemisphere of the sky, containing Proxima Centauri, the closest star to the Sun. Centaurus contains Omega Centauri, a bright globular cluster of hundreds of thousands of stars, 17,000 light years away. Also in the constellation lies Centaurus A, a galaxy that emits radio waves.

Cepheid variables are types of stars that vary regularly in light output as they alternately swell up and contract in size. They are giant stars, dozens of times larger than the Sun and hundreds or thousands of times as bright. Their brightness is directly related to how long they take to vary – the brightest Cepheid taking the longest period to vary. By observing its period of pulsation, an astronomer can calculate the brightness of a Cepheid, and hence estimate its distance. Cepheids are important distance indicators in astronomy. They take their name from Delta Cephei, the first star of the type to be discovered.

Chromosphere is a layer of glowing hydrogen gas, 16,000 km deep, above the visible surface (PHOTOSPHERE) of the Sun. It gets its name which means colour sphere, from its strong red colour as seen at eclipses when the Moon blocks out light from the far brighter photosphere.

Coal Sack is a dark cloud of dust and gas in the constellation Crux. The Coal Sack is 400 light years away, and contains enough material to make a cluster of hundreds of stars.

Constellations are patterns of stars in the sky. The Greeks named many constellations after their mythological heroes. Others have been added since, so that a total of 88 constellations can now be identified.

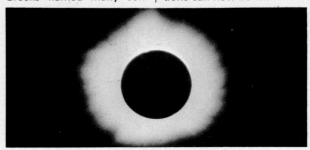

Solar eclipse showing corona

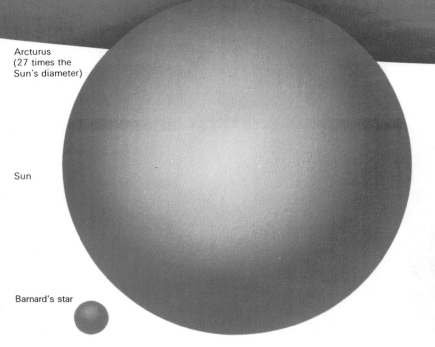

Arcturus
(27 times the
Sun's diameter)

Sun

Barnard's star

Features of the southern skies

The constellations of the ZODIAC are those constellations through which the Sun passes on its apparent annual journey through the stars. The path that it follows is called the ECLIPTIC, and the 12 constellations of the zodiac that lie along it are ARIES, TAURUS, GEMINI, CANCER, LEO, VIRGO, LIBRA, SCORPIUS, SAGITTARIUS, CAPRICORNUS, AQUARIUS and PISCES. These constellations are to be seen in the northern skies as well as the southern skies, for the ecliptic lies in the equatorial region midway between the two celestial poles (the points above the poles on Earth). The most distinctive of them is Scorpius, the Scorpion, which has a line of stars curved like a scorpion's tail.

Another distinctive constellation that is to be seen in both northern and southern skies (depending on the time of year) is ORION. It can be recognized immediately by the line of three bright stars that make up Orion's belt. The rest of the constellation is marked by four bright stars, two on either side of the belt. The line of Orion's belt points to the brightest star in the heavens, SIRIUS, and also near the belt is the Great Nebula, a mass of glowing gas visible to the naked eye as a misty patch.

The southern skies cannot really claim as their own as many bright constellations as the northern skies, but there are several interesting features. The most prominent constellation is CENTAURUS, the Centaur. It lies in the Milky Way and contains the nearest bright star to Earth, Alpha Centauri. Proxima Centauri, a companion star too faint to be seen by the naked eye, is in fact the closest star to us (apart from the Sun). It is 4.28 light years away.

Beside Centaurus is the best-known southern constellation, the Southern Cross or CRUX, which resembles a tiny cross in the sky. Nearby is the constellation of Carina, the Keel, which contains

Above: The comparative sizes of 3 stars. The Sun is an average kind of star, while Arcturus is a red giant 27 times its diameter and Barnard's star is a red dwarf star, less than twice the size of the planet Jupiter. The Earth is less than 1% of the size of the Sun and, on this scale, would be smaller than a millimetre across.
Right: The Lagoon nebula in the constellation Sagittarius. It is a mass of gas containing hot stars that produce radiation which causes the gas to glow. Within the cloud, new stars may form. The nebula is nearly 30 light years across.

Corona is the faint outer atmosphere of the Sun, seen as a pearly-white halo at a total eclipse when the Moon blocks off the blinding light of the Sun's disc. The corona consists of thin gas, boiled off from the surface of the Sun, which drifts outwards to eventually form the SOLAR WIND.
Crab nebula is a patch of glowing gas in the constellation Taurus, the remains of a star seen to explode as a SUPERNOVA by Oriental astronomers in 1054. At the centre of the Crab nebula is a flashing PULSAR, the tiny core

of the star that exploded. The Crab nebula is 6,300 light years away.
Crux, the Southern Cross, is the smallest constellation in the sky. It lies near the south celestial pole in a rich area of the Milky Way. Crux contains the dark COAL SACK nebula, as well as the 4 bright stars that mark the famous cross shape.
Cygnus, the Swan, is a prominent constellation of the Northern Hemisphere of the sky, sometimes called the Northern Cross because of its distinctive shape. Its brightest star is Deneb. Al-

berio, or Beta Cygni, is a double star at the Swan's head. The star 61 Cygni was the first to have its PARALLAX measured. Cygnus contains

Gemini

the North American nebula, a glowing cloud of gas in the Milky Way, and the Veil nebula, the remains of an ancient SUPERNOVA explosion. Cygnus X-1 is an X-ray emitting double star which may contain the first identifiable black hole.

D **Declination** is a sky coordinate, the equivalent of latitude on Earth. An object at declination + 90° is at the north celestial pole (overhead at the Earth's north pole), while declination /0° marks the celestial equator.

E **Eclipsing binaries** are pairs of stars that orbit each other, and where one star periodically passes in front of the other as seen from Earth. Such eclipses cause the total light we see to dim, so that the star's brightness seems to vary. The first eclipsing binary to be discovered was Algol.
Ecliptic is the apparent path of the Sun in front of the stars each year, which is of course actually caused by the Earth orbiting the Sun. It is given the name ecliptic because only when the Sun and Moon are near this line

Below: The Great Nebula in Orion. It can be seen with the naked eye as a misty patch south of Orion's belt.

Below: The famous star cluster known as the Pleiades, or Seven Sisters, because 7 stars can be seen with the naked eye. In fact, the cluster contains about 200 stars. The misty patches around the stars are clouds of dust that shine by reflecting the light of the stars within them.

the second brightest star in the sky, CANOPUS. The other main features of interest are the Magellanic Clouds, patches of light near the celestial pole. These clouds are in fact the nearest galaxies to our own.

The Sun

Our Sun is a very average type of star. Like most other stars, it is a globe of hot, glowing gas that shines steadily, pouring out heat and light into space. It is neither very large nor very small for a star. Around it orbit the planets of the Solar System, including the Earth. The planets are globes of rock that are lit and warmed by radiation from the Sun. The Sun is not exceptional in possessing a family of planets. Although they are too small to be seen in telescopes, planets have been detected around some of the nearer stars.

The Sun is the kind of star that astronomers call a yellow dwarf, but being so much nearer to us than any other star, it appears particularly impressive. It is almost 150 million kilometres from the Earth, a distance that its light travels in eight minutes. By comparison, light from the next nearest star takes 4.28 years to reach us.

can eclipses occur.

F **Flares** are eruptions of brightness from the surface of the Sun, usually associated with SUNSPOTS. Flares eject high-energy radiation into space, which can cause radio blackouts on Earth and upper-atmosphere displays known as aurorae.

G **Gemini,** the Twins, is a constellation of the zodiac in the Northern Hemisphere of the sky. The two brightest stars in Gemini are Castor and Pollux. The Sun passes through the con-stellation from late June to late July. A major meteor shower, the Geminids, radi-ates from the constellation in December each year.
Great Bear is the popular name for URSA MAJOR in the sky's Northern Hemisphere.

H **Hercules** is a constella-tion of the Northern Hemisphere of the sky, named after the Greek mythological hero. One of the constellation's most im-portant features is the globu-lar cluster called M13, 22,500 light years away, containing 300,000 stars. Alpha Her-culis, also called Ras Algethi, is a red giant star, about 500 times the Sun's diameter.
Hertzsprung-Russell diag-ram is a graph on which the temperature of a star is plotted against its bright-ness. The graph is named after the Danish astronomer Ejnar Hertzsprung (1873–1967) and the Ameri-can astronomer Henry Norris Russell (1877–1957) who independently came up with the same idea in 1911 and 1913 respectively. The diagram is useful for reveal-ing whether a star is a giant or a dwarf, and what stage it has reached in its evolution. Most stars, including the Sun, lie on a band known as the MAIN SEQUENCE. A star's position on the diagram also reveals its ABSOLUTE MAG-NITUDE.
Hyades is a 'V'-shaped clus-ter of about 200 stars in the constellation Taurus. The

Leo and Cancer

The Sun has a diameter of 1,390,000 kilometres — 109 times the diameter of the Earth. Its mass is 333,000 times the Earth's mass and its volume is over a million times greater.

At the Sun's core, where hydrogen is converted into helium to produce energy, the temperature is about 15 million°C and the pressure is 400,000 million times that of the Earth's atmosphere. Such conditions are necessary to sustain thermonuclear fusion at the heart of a star. Heat is

Below: The Sun is a globe of hot gases. The temperature at its surface is about 6,000°C. From time to time dark patches appear on the surface, slowly moving around the Sun as it rotates. These patches are called sunspots (*below left*), and consist of cooler regions of gas probably produced by magnetic disturbances.

Sunspots usually form in groups, which may be as much as 200,000 kilometres across. Associated with sunspots are solar flares (*below centre*), intense eruptions in which streams of high-energy radiation and particles are ejected from the Sun. Prominences may also occur (*below right*). These are huge columns of

glowing gas that arch up into the Sun's atmosphere or out into space.

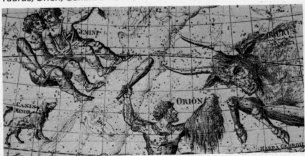

Hyades are the nearest dense star cluster to the Earth, about 150 light years away; much of our knowledge about stars comes from its study. The stars of the Hyades were born about 500 million years ago and are relatively young.

L **Leo,** the Lion is a constellation of the zodiac, lying in the Northern Hemisphere of the sky. The Sun passes through the constellation from mid-August to mid-September. Leo's brightest star is Regulus. Each November, the Leonid

meteor shower radiates from the constellation.
Libra, the Scales, is an inconspicuous constellation of the zodiac, lying in the Southern Hemisphere. The Sun passes through the constellation during November.
Light years, *see page 9.*
Lyra, the Lyre, is a small but prominent constellation of the northern sky, whose brightest star is Vega. Vega Lyrae is a famous ECLIPSING BINARY star, the components of which are distorted by each other's gravity and from which gas spirals into space. Epsilon Lyrae, some-

times called the 'double double', is a group of four stars connected by gravity. The Ring nebula in Lyra is a famous PLANETARY NEBULA.

M **Magnitude** is a measure of a star's brightness. Each magnitude step corresponds to a brightness difference of just over 2.5 times, so that a sixth magnitude star (the faintest visible to the naked eye) is 100 times fainter than a first magnitude star. Objects brighter than magnitude 0 are given negative magnitudes: for instance, Sirius

Taurus, Orion, Gemini and Canis Minor

is magnitude −1.47, Venus at its brightest is −4.3, and the Sun is −26.5 (these are all apparent magnitudes, the

brightness as seen from Earth). The faintest objects detected in telescopes are around magnitude +25.

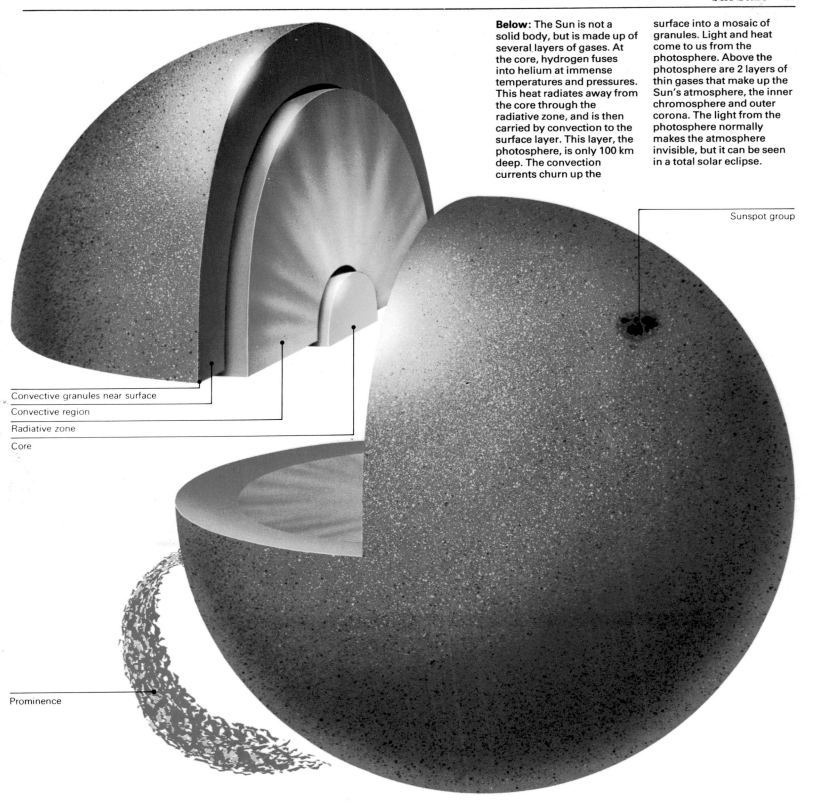

Below: The Sun is not a solid body, but is made up of several layers of gases. At the core, hydrogen fuses into helium at immense temperatures and pressures. This heat radiates away from the core through the radiative zone, and is then carried by convection to the surface layer. This layer, the photosphere, is only 100 km deep. The convection currents churn up the surface into a mosaic of granules. Light and heat come to us from the photosphere. Above the photosphere are 2 layers of thin gases that make up the Sun's atmosphere, the inner chromosphere and outer corona. The light from the photosphere normally makes the atmosphere invisible, but it can be seen in a total solar eclipse.

Sunspot group

Convective granules near surface

Convective region

Radiative zone

Core

Prominence

Main sequence is the stable middle-age of a star's evolution, and the stage at which the Sun and most other stars are. Main-sequence stars are burning hydrogen into helium at their centres to create energy. When that hydrogen store is used up, the star evolves out of the main sequence to become a giant.

N **Nebula** is a mass of dust and gas in the galaxy. Some nebulae are bright, shining by the light of stars embedded within them; the ORION nebula is one such example. Others are dark, like the COAL SACK.

Neutron stars are small dense stars believed to mark the end point of the evolution of stars more massive than the Sun. A neutron star is only about 15 km across, yet contains as much matter as our Sun, compressed so tightly that a thimbleful of neutron star material would weigh thousands of millions of tonnes. The flashing radio sources called PULSARS are actually neutron stars.

Nova is an exploding star, which flares up in brightness by 10,000 times or more in a day before fading again over a period of weeks or months. Novae are believed to be double star systems in which gas flows from one star onto a companion white dwarf. This gas ignites and is thrown off from the white dwarf, causing the eruption of brightness. A star is not devastated by a nova explosion, so that novae can occur again.

O **Orion** is a large and prominent constellation of the equatorial region of the sky, named after a Greek mythological hunter. Its brightest stars are BETELGEUSE and Rigel. A distinctive line of 3 stars represents Orion's belt. Many of the stars in Orion are relatively young because Orion marks an area of star formation, notably in the Orion nebula.

Orion nebula is a cloud of gas and dust about 1,500

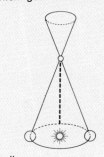

Parallax

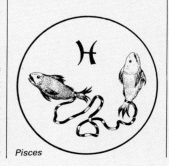

Pisces

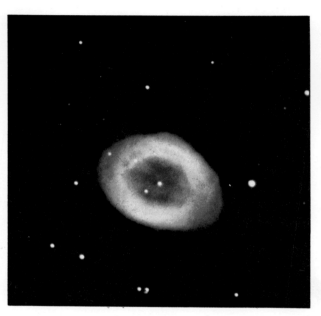

The Evolution of the Sun
1. The Sun, and the rest of the Solar System, began as a cloud of dust and hydrogen gas floating in space. Neither the Sun nor the planets existed.

2. More than 5,000 million years ago, the cloud began to shrink as gravity pulled the particles in the cloud together. As the cloud condensed, it started to spin and heat up.

3. At the centre of the cloud, heating was most intense. A central hot body began to emerge around which the outer part of the cloud revolved.

4. The central body became hotter as it contracted. The outer part of the cloud began to condense into separate smaller clouds.

5. The particles in the outer clouds began to come together to form separate bodies. More particles may have come from the central body or from space.

Stars vary widely in brightness and in colour. The brightness of a star is called its MAGNITUDE, and there are two ways of measuring it. The APPARENT MAGNITUDE is the brightness of a star as it appears to us on Earth. This depends on its distance from the Earth, for nearby stars appear brighter than distant stars of the same actual brightness. The magnitude of a heavenly body is denoted by a number that simply shows how much brighter or dimmer it is than another. The brighter the object, the smaller the numerical value of the magnitude. This scale was introduced by the Greek astronomer Hipparchus over 2,000 years ago. ANTARES has an apparent magnitude of 1.00, and the naked eye can see stars down to a magnitude of about 6. The most powerful telescopes can detect stars with a magnitude of 25, which are about 4,000 million times dimmer than stars of the first magnitude. Several stars are brighter than Antares, and the scale extends into minus figures. Thus, Sirius has an apparent magnitude of −1.47 and the planet Venus may get as bright as magnitude −4.3. The Sun's apparent magnitude is −26.5.

Astronomers can also find out the actual brightness of a star, which is of more value than the apparent magnitude. This ABSOLUTE MAGNITUDE is defined as the apparent magnitude of the star if it were at a distance of 10 parsecs (32.6 light years) from Earth. On this scale, the brightest known star is believed to be one in the Large Magellanic Cloud with an absolute magnitude of −8.9. By comparison, Sirius has an absolute magnitude of 1.41 and the Sun is relatively dim, at 4.83.

Left: The Ring nebula in Lyra is a well-known planetary nebula. It probably consists of a shell of hot gas surrounding a star that is collapsing inwards after shedding its outer layers towards the end of its life.

Even with the naked eye, some stars appear to be of different colours, In fact, stars do vary in colour from red, through yellow and white to blue. The colour depends on the temperature of the star, cooler stars being red and the hottest stars blue. The Sun is of average temperature, and is therefore yellow. Although the Sun shines steadily, many stars do not. These vary in brightness, either in a regular cycle or irregularly. There are two important kinds of VARIABLE STARS. The first are ECLIPSING BINARIES, or double stars — pairs of stars that revolve around one another. Many stars in fact consist of pairs or even greater numbers in association with each other. In an eclipsing binary, one star passes in

14. The Sun will become a white dwarf star, about the size of the Earth but still producing some heat.
15. Eventually, energy production must cease and the Sun will finally die, becoming a cold, dark body.

light years away in which new stars are forming. Part of the nebula is visible to the naked eye as a fuzzy, glowing patch marking the sword of Orion. Newly-formed stars at the heart of the nebula make it glow, but radio astronomers have also detected an even larger dark cloud behind the visible portion in which stars are still being born.

P Parallax is an object's shift in position against a distant background as seen from 2 different places. Stars show a slight parallax

shift when viewed from opposite sides of the Earth's orbit. The amount of shift depends on the star's distance, the closest stars showing the greatest shift. By measuring a star's parallax, astronomers can work out its distance. Only stars closer than about 100 light years show sufficient parallax to be measured accurately. The star 61 Cygni was the first to have its parallax measured in 1838 by the German astronomer Friedrich Wilhelm Bessel (1794–1846).
Parsec is a measure of dis-

tance in astronomy. It is the distance at which a star would show a PARALLAX of 1 second of arc. A parsec is equal to approximately 3.26 light years.
Perseus is a prominent constellation of the Northern Hemisphere of the sky, named after a Greek

Hewish and Bell (pulsar discoverers)

mythological hero. Perseus lies in a dense part of the Milky Way. Its most famous star is Algol. Every August, a rich shower of meteors, the Perseids, radiates from the constellation.
Photosphere is the brilliant surface of the Sun; its name means 'sphere of light'. It is a layer of glowing gas at a temperature of about 6,000°C. The photosphere is broken up by convection cells of hot gas, called granules, each about the size of Britain. Cooler areas of the photosphere are called SUNSPOTS.

9. Then the supply of hydrogen fuel in the Sun will begin to run low. The core will begin to contract.

6. Fusion reactions, in which hydrogen changes into helium, began in the central body as contraction produced great heat and pressure. The outer clouds, too small to create fusion, continued to condense into solid bodies.

7. The central body became a star – the Sun – as the energy expansion caused by fusion balanced the contraction caused by gravity. It began to shine steadily about 5,000 million years ago. The planets formed at about the same time.

8. The Sun will continue to shine steadily for another 5,000 million years. The Solar System should remain unchanged for this time.

10. Contraction will produce more energy, causing the Sun to stop contracting and expand.

11. The Sun will begin to swell and become a red giant star, swallowing up the inner planets, including the Earth.

12. Helium formed by fusion during the Sun's lifetime will now begin to burn, making the Sun expand even more.

13. The Sun will become unstable as its helium runs out. It will shed its outer layers and the remainder will collapse inwards.

Pisces, the Fishes, is a constellation of the zodiac, lying in the equatorial region of the sky. The Sun passes through the constellation from mid-March to mid-April, and is in Pisces when it moves north across the celestial equator, marking the start of the Northern Hemisphere spring (the vernal equinox).

Planetary nebula is a shell of gas that in a small telescope shows a disk like that of a planet, hence the name. In fact they have nothing to do with planets, but are believed to be the outer layers of former RED GIANT stars which have drifted off into space, leaving the star's core as a WHITE DWARF.

Pleiades are a cluster of about 200 stars, 415 light years away in the constellation Taurus. It is possible to see 6 or 7 with the naked eye, hence the group's popular name, the seven sisters. The Pleiades are relatively young, the youngest having formed within the past few million years.

Plough is the popular name for the most prominent part of the constellation URSA MAJOR in the northern sky.

Pulsar in Crab nebula

Polaris, the Pole Star, is the brightest star in the constellation Ursa Minor. By chance, it lies about 1° from the celestial North Pole.

Prominences are clouds of hot gas projecting from the surface of the Sun, associated with strong magnetic fields. Some prominences are shaped like arches, and can last for weeks or months, while others, often associated with FLARES, shoot off into space at up to 1,000 km per second.

Proper motion is a slight change in a star's position over a period of time caused by its movement through space. Proper motions of stars cannot be seen by the naked eye, but can be measured on large-scale photographs taken by telescopes over many years. The proper motion of stars will eventually change the familiar constellation patterns. The star with the greatest proper motion, BARNARD'S STAR, over a period of 180 years, changes its position by as much as the apparent diameter of the Moon.

Pulsar is a rapidly pulsating radio source, believed to be a spinning neutron star

front of the other as it revolves around it, temporarily cutting off its light. ALGOL is a well-known eclipsing binary which fades in brightness every three days.

The other main kind of variable star is called a CEPHEID VARIABLE because the first one was discovered in the constellation Cepheus. Cepheid variables change brightness in regular cycles lasting from a few hours up to several weeks. The variations occur because the stars regularly expand and contract. The absolute magnitude of a Cepheid variable can be calculated by the length of its cycle. By finding the absolute magnitude in this way and then measuring its apparent magnitude, the distance of the star can be plotted. This is a very useful method for finding the distances of stars far away in our galaxy. Furthermore, the distances of nearby galaxies can be found by detecting any Cepheid variables they may contain.

The life of a star

Whatever their differences, all stars begin life in the same way.

The space that lies between the stars is not entirely empty. Thin clouds of dust and hydrogen gas float here and there, sometimes so thickly as to obscure our view of the stars that lie beyond. We cannot see the centre of our galaxy for this reason. However, in places the clouds begin to condense and become thicker and thicker. This is because the particles in the cloud are pulled towards each other by their gravity. As the cloud condenses, it begins to heat up. Over millions of years, it collapses into a ball of hydrogen gas that becomes so hot that thermonuclear reactions start and the hydrogen begins to turn into helium. The resulting heat stops the ball of gas from collapsing any further, and it begins to shine steadily. The star is born.

Astronomers believe that they can see this process of star formation going on in such nebulae as the ORION NEBULA. Some of these clouds of gas have stars embedded within them and are flaring with light. The stars are thought to be young stars and other stars are probably forming around them.

The Sun is a typical star. It began life like other stars and will end like most of them. The Sun has been shining steadily for about 5,000 million years and it will continue to shine for

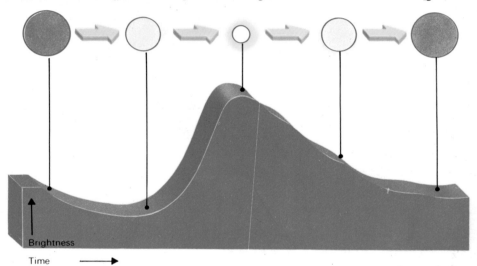

Brightness

Time →

| Star at its largest | Star contracts | Star at its smallest | Star expands | Star at its largest |

Above: An eclipsing binary is a double star system that makes regular but sudden changes in brightness. This happens as one star passes in front of, or behind, the other in its orbit around it as seen from Earth. In this case, the smaller star is the brightest, and the greatest variation occurs as it passes behind the other star. In the case of the best-known eclipsing binary, Algol, the central star is the brightest.

Brightness

Time →

Above: A cepheid variable is another kind of variable star. The pattern of variation is different to that of an eclipsing binary in that brightness regularly comes to a sudden peak and then tails off. Cepheid variables vary in brightness because they regularly expand and contract, being at their brightest when smallest. The cycle may last from less than a day to several weeks, depending on the overall brightness.

---

which emits a beam of radiation like a lighthouse beam. Pulsars were discovered in 1967 by Cambridge radio astronomer Anthony Hewish and his assistant Jocelyn Bell. About 150 are now known. The fastest pulsar, at the heart of the Crab nebula, pulsates 30 times a second and can also be seen flashing optically. The slowest pulsars pulsate once every 3 seconds or so.

R **Red dwarfs** are cool, faint stars, less massive than the Sun. They are probably the most abundant stars in our galaxy, but are difficult to see because they are so faint. Even the closest red dwarfs Proxima Centauri and Barnard's star, are invisible without a telescope.

Sagittarius

Red giants are stars bigger than the Sun, with a cool surface. They are believed to form when stars swell up in size at the end of their life. The Sun is expected to become a red giant, like Arcturus, in about 5,000 million years. Stars which become dozens or even hundreds of times the diameter of the Sun are called super giants.

Right ascension is a co-ordinate in the sky equivalent to longitude on Earth. It is measured in hours, minutes, and seconds, from 0h to 24h. The starting point of right ascension is where the Sun moves north across the celestial equator. This marks the start of the vernal equinox, when the Sun is passing through Pisces.

S **Sagittarius,** the Archer, is a constellation of the zodiac lying in the Southern Hemisphere of the sky. The Sun passes through the constellation briefly at the end of November. One of its stars was the first to be recognized as a double star. **Scorpius,** the Scorpion, is a constellation of the Southern Hemisphere of the sky.

Its brightest star is Antares, a double star. The Sun passes through the constellation between the end of October and mid-November.

Solar System and zodiac

Above: The Crab nebula in Taurus is the remains of a star that exploded and became a supernova in 1054. Oriental astronomers witnessed the event.

another 5,000 million years, but then it will run out of its hydrogen fuel. The core, exhausted of fuel, will contract and produce more energy as it does so. This energy will cause the Sun to swell into a red giant star, engulfing the Earth. It will then begin to shed its outer layers, producing a large cloud known as a PLANETARY NEBULA. Finally, the remainder will cool and shrink to the size of a planet, becoming a white dwarf star so dense that a piece of it the size of a sugar lump would weigh a tonne.

This is a dramatic ending — especially where our own planet is concerned — but some stars really go out with a bang. Strangely enough, those stars larger than the Sun do not last longer. Large stars burn at a higher temperature and use up their fuel quickly, living for as little as a few million years. As the fuel is exhausted, the star becomes unstable and blows up in a vast explosion, becoming a SUPERNOVA. Such catastrophes are not frequent, occurring only once every century or so in our galaxy. But they have not gone unnoticed in the past, for the exploding star becomes so bright that it is visible by day. The CRAB NEBULA is the remains of a supernova that was recorded in 1054. Tycho BRAHE (*see page 5*) made his name by observing another in 1572, calling it *nova stella*, meaning new star. Astronomers now use the term NOVA to describe a star that suddenly becomes very bright as a result of a convulsion insufficient to devastate the star. A supernova is a star that blows itself to pieces.

Hydrogen and helium are different chemical elements. Everything in the Universe, both living and non-living, is composed of chemical elements, of which 105 are known in all. As stars get hotter and hotter towards the end of their lives, other elements are formed from the helium already produced. When the star sheds its outer layers or explodes in a supernova, it ejects these elements out into space as gas and dust. This gas and dust may then eventually become part of a cloud that condenses into a new star. In this way, atoms of the various elements may be recycled throughout the Universe. Some of the atoms that make up our bodies and our world may have been born in the death-throes of far-off stars.

Pulsars and black holes

A supernova does not mean the total end of a star; in fact, it marks the beginning of a final

---

Sirius is the brightest star in the night sky. It is a hot, white star, 1.75 times the Sun's diameter, and lies 8.7 light years away. Sirius is orbited every 50 years by a white dwarf companion star.

Solar cycle is the term used for the approximate 11-year variation in activity on the Sun. The number of sunspots, flares, and prominences varies during each 11-year cycle. However, in each successive cycle the north and south magnetic poles of the Sun interchange, so it may be said that the complete solar cycle

takes 22 years. The last solar maximum was in 1969–70, and the next is expected in 1982.

Solar winds are streams of atomic particles from the Sun flowing outwards through the Solar System. The solar wind is an extension of the Sun's CORONA.

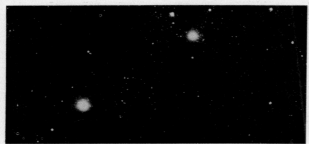

Southern Cross and Coal Sack

The solar wind is detected flowing past the Earth, so it can be said that we lie in the outer regions of the Sun's corona.

Southern Cross is the popular name for the constellation CRUX.

Sun is our parent star, nearly 150 million km away. The Sun is an average star, 1.4 million km in diameter, made up mostly of hydrogen and helium gas. The Sun generates light and heat by nuclear reactions at its centre. It is orbited by a family of 9 planets, including the Earth.

Sunspots are cooler areas on the Sun's surface, the PHOTOSPHERE. They appear dark by contrast with their more brilliant surroundings, but they are themselves quite hot, about 4,500°C. Sunspots are associated with magnetic fields on the Sun, which apparently block the flow of heat through the photosphere to cause the spot.

Supernova is the brilliant explosion of a massive star at the end of its life. In a supernova the star flares up by as much as several million times its normal bright-

stage of existence that has only been discovered recently. In 1967, radio astronomers at the Mullard Observatory in Cambridge, England, were puzzled by a new signal that consisted of regular rapid pulses of radio waves. It came from deep space, and the astronomers wondered whether it was a message from the stars. But the signal did not vary, showing it to be natural in origin. The source was called a PULSAR. Others were soon discovered and a few were also found to produce visible light, flashing on and off like a beacon in space. One of these visual pulsars was found at the heart of the Crab nebula.

Astronomers believe pulsars to be NEUTRON STARS. As a huge star collapses and explodes in a supernova, the atoms at its core are forced together so that the electrons enter the nuclei of the atoms and produce neutrons. The star ends up as a ball of neutrons only a few kilometres across but so dense that a piece the size of a sugar lump would weigh millions of tonnes.

As it collapses, the star turns faster and faster so that the resulting neutron star spins extremely rapidly — 30 times a second in the case of the Crab nebula pulsar. It radiates beams of energy, such as light and radio waves, and these beams sweep round like the beam of a lighthouse. It thus appears to flash on and off.

Even this may not be the end. As a star collapses and its matter is forced into a state of extreme compression, its force of gravity increases. If it were to continue its collapse past the stage of a neutron star, the gravitational field around the collapsed body could become so strong that nothing would be able to escape from it. Even light waves would not be able to leave and the body would become totally black. Any object approaching the body would be pulled down to its surface, never to escape again. Such a body is known as a BLACK HOLE. The existence of black holes has only been predicted from theory; there is no hard evidence to show that they exist. Indeed, how could such a body produce any sign of its existence? Theory predicts that matter such as gas falling into a black hole produces energy as it accelerates to its doom. A rapidly varying X-ray source discovered in the constellation Cygnus could be a black hole. The enormous energy output of QUASARS (*see page 12*) could be explained by assuming that black holes lie at their hearts.

Formation of a black hole
1. Light leaves a normal star hardly affected by its field of gravity. In the case of an average star like the Sun, the light is yellow in colour.

2. As a star collapses at the end of its life, its pull of gravity increases enormously. This produces a large red shift in the star's light, making it appear redder.

3. If the star collapses and becomes a black hole, gravity increases to the point at which light cannot escape. Light rays passing nearby are deflected by gravity and drawn into the black hole. It appears totally black.

---

ness. The outer layers of the star are thrown off, forming an object like the Crab nebula, while the star's core may become a neutron star or BLACK HOLE.

T **Taurus,** the Bull, is a large constellation of the zodiac in the Northern Hemisphere of the sky. The Sun passes through the constellation from mid-May to late June. The brightest star in Taurus is Aldebaran. The constellation contains the Hyades and Pleiades star clusters, as well as the Crab nebula.

Sundial

U **Ursa Major,** the Great Bear, is a prominent constellation of the Northern Hemisphere of the sky. Its 7 brightest stars make up a saucepan shape often referred to as the Plough. Two stars of the bowl point to POLARIS. The second star along the handle, called Mizar, has a faint companion, called Alcor, visible with keen eyesight or through binoculars.
Ursa Minor, the Lesser Bear, is a constellation at the North Pole of the sky. The brightest star is Polaris.

V **Variable stars** are stars whose light output changes. Some of the stars change in size, such as CEPHEID VARIABLES, but others are close double stars that periodically eclipse each other (ECLIPSING BINARIES). In 1975 over 25,000 variable stars were catalogued in our galaxy.
Virgo, the Virgin, is a constellation of the zodiac lying in the equatorial region of the sky. The Sun passes through the constellation from mid-September to early November. The brightest star in Virgo is Spica.

W **White dwarf** is a small, hot star believed to mark the end point of the evolution of a star such as the Sun. A white dwarf is about the size of Earth, but contains as much matter as the Sun, compacted so densely that a thimbleful would weigh a tonne or more. White dwarfs are so faint that even the nearest, around Sirius and Procyon, are only visible with the aid of a telescope.

Z **Zodiac** is the band of 12 constellations which the Sun passes in front of during the year. The signs used by astrologers do not correspond with the constellations of the same name.

Space probes are supplying information about the planets and other bodies in our Solar System. It now seems unlikely that life – as we know it – exists on other planets, but distant Solar Systems, with inhabited planets, may exist.

The Solar System

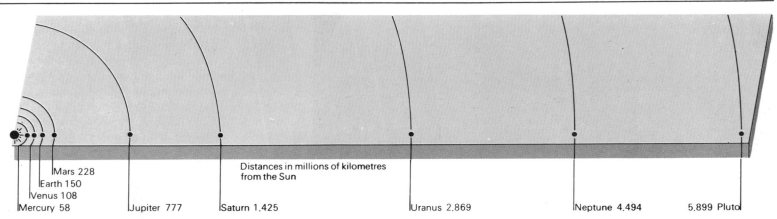

Distances in millions of kilometres from the Sun

Mercury 58 | Venus 108 | Earth 150 | Mars 228 | Jupiter 777 | Saturn 1,425 | Uranus 2,869 | Neptune 4,494 | 5,899 Pluto

The Sun has a great family of PLANETS and their moons, as well as ASTEROIDS, COMETS, debris that we see as METEORS and a few man-made spacecraft within its field of gravity. All these bodies make up the SOLAR SYSTEM, and they all move in ORBITS determined by the Sun's gravity. Planets also have moons that revolve around them as they orbit the Sun, and the paths of asteroids, comets and spacecraft may be influenced by any planets they approach.

The asteroids, comets and meteors of the Solar System all move round the Sun in the same direction. If viewed from a point in space far above the North Pole on Earth, this motion would appear to be anti-clockwise. With the exception of one, all the planets also spin in an anti-clockwise direction; the exception is Venus, which has a clockwise or RETROGRADE rotation. The planets move in almost the same plane as the Earth. Only Mercury and Pluto depart significantly with INCLINATIONS of 7° and 17° respectively. The planets move around the Sun in ellipses so that their distance from the Sun varies. Most of the orbits are almost circular, so that there is little difference between APHELION and PERIHELION, the farthest and nearest orbital points. However, the orbit of Pluto – normally the outermost planet of the Solar System – is so

Above: The planets of the Solar System with their average distances from the Sun to scale.

eccentric that at perihelion, it is closer to the Sun than Neptune. The Sun's grip really extends much farther out into space; some comets have such eccentric orbits that their aphelion may be halfway to the next star, a distance of about 20 million million kilometres.

Moons orbit around most of the planets, varying in number from Earth's single moon to Jupiter's 13 or possibly 14 companions. Only Mercury, Venus and Pluto have no moons. Most of the moons orbit their planets in an anti-clockwise direction, moving in a plane around the planet's equator (at right angles to its axis). Moons are a planet's natural satellites. Earth, our Moon and Mars also have man-made satellites in orbit around them, and other spacecraft flying between the planets become man-made satellites of the Sun.

The origin of the Solar System
The Sun formed as a cloud of hydrogen gas and dust condensed into a ball that heated up and began to shine about 5,000 million years ago. The oldest rock found on Earth is 3,800 million years old, but meteorites and Moon rocks have been dated and found to be as old as 4,600 million years. Geologists believe that the Earth is this age too, but no rocks as old as this have been

Reference

John Couch Adams

the position predicted by both men.

A **Adams,** John Couch (1819–92) was an English astronomer who calculated the existence of the planet Neptune. In 1843 he began to investigate the erratic motion of the planet Uranus, which he believed could be attributed to the gravitational pull of an unseen planet. Similar calculations were made independently in France by Urbain LEVERRIER. Neptune was eventually discovered close to

Albedo is the proportion of light that a body reflects. A perfectly white body reflects all the light that hits it, and thus has an albedo of 1. A totally black body reflects no light, so its albedo is 0. A body's albedo is a guide to the nature of its surface.
Aphelion is the farthest orbital point that a body reaches from the Sun; it is the opposite point to PERIHELION. For instance, the Earth at aphelion, which it reaches in early July, is 5 million km further from the Sun than at perihelion, which occurs in early January.
Apogee is the farthest point

that an object, such as the Moon or a satellite, orbits around the Earth. It is opposite the PERIGEE.
Asteroid is a small lump of rock or metal orbiting the Sun; an alternative name is 'minor planet'. Most asteroids orbit in a belt between Mars and Jupiter, but some have more eccentric orbits that take them in past the Earth or out past Jupiter. A small group of asteroids, called the TROJANS, move in the same orbit as Jupiter. At present about 2,500 asteroids have had their orbits calculated. The asteroids

probably represent rubble left over from the formation of the Solar System. Their total mass is estimated to be only a few per cent of the Moon's mass so they are unlikely to be the shattered remains of a former planet, as one theory has suggested.
Astronomical unit is the average distance between the Sun and Earth. It is equivalent to 149,597,910 km.
Aurora is a glowing light display in the upper atmosphere, caused by the impact of atomic particles ejected

More information was obtained by bouncing radar signals off the planet. These showed that the surface was probably covered with craters, like the Moon, and they also reported that the planet rotates once every 59 days and not 88 days, as previously thought. The period of rotation is exactly two-thirds of Mercury's year (the time it takes to revolve once around the Sun). For this reason, a day on Mercury from one sunrise to the next lasts 176 Earth-days or two of Mercury's years! Conditions are therefore extreme on Mercury's surface, where there is virtually no atmosphere to protect it. The Sun bakes the surface during the year-long daytime, producing a temperature of 325°C. And during the equally long night, it plunges to −170°C.

Man's first good look at Mercury came in March 1974, when the American space probe MARINER 10 flew past and sent pictures and

discovered, possibly because they have been destroyed in geological changes.

Most astronomers hold that the planets formed as clouds that were made up mainly of light gases with a small amount of heavier gas which sank towards the centres of the clouds to form dense cores. The planetary moons probably formed in the same way and went into orbit around their planets instead of the Sun. However, views differ as to the origin of the material making up the planets and moons. It may have been the gas left over from the cloud that condensed to form the Sun; or the Sun, once it had formed, pulled in more material from space which condensed to form the planets and moons. Either way, the formation of planets does not seem to be exceptional as bodies have been detected in orbit around nearby stars.

Mercury

MERCURY is the closest planet to the Sun, and has only one-third of the diameter of the Earth. Being near the Sun and being so small, it is difficult to get a good view of it in a telescope. Furthermore, being nearer the Sun than the Earth, Mercury shows phases like the Moon. It can only be seen fully illuminated when it is on the opposite side of the Sun. When it is nearest to us, we see only the shaded side. With telescopes, astronomers made maps of Mercury, but they showed little more than vague bright and dark areas.

Above: The first close views of Mercury were obtained by the American space probe Mariner 10 in 1974. They showed a world remarkably like the Moon in appearance.

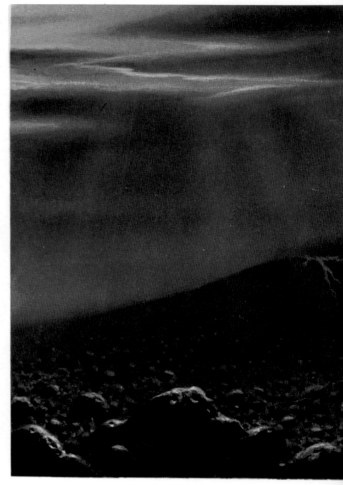

Right: An artist's impression of the Russian space probe Venus 10, which landed on the surface of Venus in 1976 and sent back the first pictures of the planet's terrain. The pictures showed a rough surface littered with rocks. This surprised scientists, because it is very hot on Venus and the clouds are laden with sulphuric acid – conditions that should rapidly wear rocks away. It therefore seems possible that there may be volcanoes on Venus producing new rocks. Radar observations indicate the presence of craters on the surface, a feature that could also be volcanic in origin.

by explosions on the Sun. The particles, mostly electrons, are channelled down towards the magnetic poles by the Earth's magnetic field, which is why aurorae are seen predominantly in polar regions.

Aurora display in the polar skies

B **Bode's law** is a series of numbers that describes the approximate relative distances of the planets from the Sun, as far as Saturn. It is named after the German astronomer Johann Elert Bode (1747–1826) who in 1772 showed that if the distance of Mercury from the Sun is divided into 4 parts and to each is added 3, 6, 12, 24, 48 and 96, the results are the relative distances of the planets out to Saturn. There was a gap at 4 + 24 = 28 on this scheme, which was later found to be filled by the asteroids. Modern as-

tronomers now attach very little significance to this law.

C **Canals** of Mars were first drawn by Giovanni SCHIAPARELLI in 1877, but reached their greatest fame through the work of Percival LOWELL at the turn of the century. Lowell drew maps of a network of canals on Mars which he believed were genuine irrigation channels dug by Martians. Most other astronomers failed to see them, and space probe photographs have conclusively proved that they do not exist. The canals

that Lowell and his followers saw must have been optical illusions.

Cassini, Jean Dominique (1625–1712) was a French astronomer, born in Italy, who in 1675 discovered the gap devoid of particles in Saturn's rings now called Cassini's division. At the Paris Observatory, where he was director from 1669, Cassini discovered the satellites of Saturn: Iapetus, Rhea, Tethys and Dione.

Ceres is the largest asteroid, 1,000 km in diameter, and the first to have been discovered in 1801 by Giuseppe

information back to Earth. The probe had flown past Venus on the way, and was thrown into an orbit that returned it to Mercury every six months. It made two more visits before its instruments gave out. The pictures show that Mercury is indeed dotted with craters like the Moon. Life is just as unlikely to exist there as on the Moon, and it is rather unlikely that man will visit the planet. However, Mercury would make an ideal place for a base to study the Sun, and an automatic station may one day be set up there.

Venus

Although VENUS comes closer to Earth than any other planet, it has long remained a mystery to us. It is not only difficult to observe, for the same reasons as is Mercury, but the whole planet is always covered with white clouds. Only when space probes began to visit the planet in 1962 did

⚪ **Mercury**

⚪ **Venus**

🌑 **Earth**

Above: The planets Mercury, Venus and Earth to scale.

Right: A view of Venus taken in ultra-violet light by the American space probe Mariner 10 in 1974 while on its way to Mercury. The pictures revealed bands of cloud moving around the planet once every 4 days. As Venus rotates very slowly, the motion of the clouds must be due to very high winds.

Jean Cassini

PIAZZI. Ceres orbits the Sun every 4.6 years at an average distance of 413.8 million km.

Comets are composed of dust and gas orbiting the Sun. Their extremely elliptical orbits take many thousands of years to complete. Some comets have shorter orbits so that they reappear at regular intervals, such as the famous HALLEY'S COMET. A dozen or more comets may be seen in a year, but only a few become bright enough to be seen by the naked eye. Even the largest comets weigh no more than a millionth of the Earth. Comets fade out and die in their old age as they lose their gas and dust.

Conjunction is the lining up of a Solar System body with the Sun and Earth. Objects closer to the Sun than the Earth are said to be at 'superior conjunction' when they are on the far side of the Sun from the Earth, and at 'inferior conjunction' when they lie between us and the Sun. Planets beyond the Earth are at conjunction when behind the Sun, since they never come between the Earth and Sun.

Craters are found on rocky bodies throughout the Solar System. The Moon and Mercury are almost completely

covered with craters, most of which are believed to have been formed by the impact of METEORITES. Mars has craters too, some of meteoric origin and some formed by volcanoes. The Earth has been largely shel-

Mariner 10's close-up of Mercury's craters

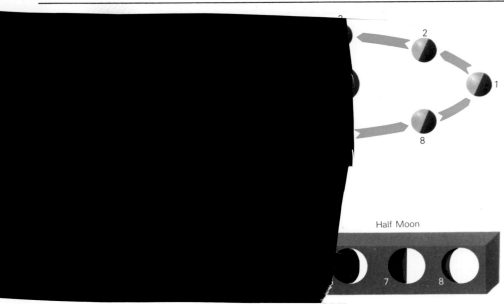

Half Moon

Earth

Moon

Above: The Earth and Moon to scale.

Above: The phases of the Moon occur because we see varying portions of the Moon lit by the Sun as it revolves around the Earth. When the Moon is completely illuminated by the Sun, we see a full moon. When we see only the shaded side it is a new moon, and inbetween comes the half moons as the moon waxes and wanes.

Below: This specimen of Moon rock was collected by the crew of Apollo 12. It comes from the Ocean of Storms and shows cavities which were formed by the expansion of vapour in a molten mass. Several Moon rocks were analyzed by the world's leading scientists.

any idea of conditions there begin to emerge. Then a series of Russian and American probes found that the temperature at the surface is a scorching 475°C – hotter than any other place in the Solar System. The heat quickly destroyed any probes that attempted to land but in 1976, pictures of the rocky surface were at last sent back by the Russian probe Venera 10. Venus should yield more of its secrets to the PIONEER probes to be launched by the United States in 1978. One will orbit Venus, studying its atmosphere and surface by radar, and another will send entry probes down to the planet. However, there is no possibility that life can exist on Venus; the planet is far too hot.

Earth

The EARTH is an unusual planet. Its distance from the Sun makes it neither too hot nor too cold for liquid water to exist. So great oceans cover much of the planet and rain-bearing clouds swirl in the atmosphere, making our world a planet that, from space, rivals Saturn in beauty. The atmosphere moderates the temperature so that extremes do not occur between night and day and it contains oxygen, making life as we know it possible. Life can also exist because the atmosphere prevents harmful rays and particles from the Sun reaching the surface. The Earth's magnetic field traps the particles in zones called VAN ALLEN BELTS high above the ground. Where the particles get past the belts above the poles, displays of light called AURORAE occur as they strike the atmosphere.

The Earth's axis is tilted at an angle of 23.5° to a line at right angles to the plane of its orbit. Like a spinning top, the Earth's rotation causes it to keep its axis pointing in the same direction relative to the stars as it orbits the Sun, apart from slight movements known as PRECESSION and NUTATION. As a result, the polar regions are alternately inclined towards and away from the Sun as the Earth revolves around it. This causes the Sun to vary its height in the sky and thus causes the seasons to occur. The passage of the Sun through the sky is marked by the EQUINOXES and SOLSTICES.

Moon

Compared with the Earth, the MOON is relatively much larger than any other satellite, having

---

tered from meteorites by its atmosphere, but there are still a number of meteorite craters on its surface, such as the famous 1-km wide crater near Flagstaff, Arizona, caused possibly 50,000 years ago.

D **Day** is the time the Earth takes to rotate once on its axis. For convenience a day is divided into 24 hours. The axial rotation period of other planets is also referred to as their 'day' although in several cases, such as Mercury, Venus, and Uranus, this is not equiva-

lent to the time between successive sunrises.
Deimos is the smaller and the more distant of the 2 moons of Mars. Deimos is a rocky, irregular-shaped body with an average diameter of about 12 km. It orbits Mars every 1.26 days at an average distance of 20,000 km from the surface. It was discovered in 1877 by Asaph Hall.

E **Earth** is our home planet, third in distance from the Sun. Its diameter is 12,742 km (average), and it orbits the Sun every

365.2422 days at an average distance of approximately 149.6 million km. The average density of the Earth is 5.5 times that of water. Its

Earth from the Moon

strong magnetic field, plus this particular density, indicates that under its outer layers of light rocks is probably a core of dense iron and nickel. The Earth has one natural satellite, the Moon, whose diameter is so large in relation to the Earth (approximately 25 per cent, a far greater proportion than the satellite of any other planet) that many people regard the Earth-Moon pairing as a double planet.
Eclipse is the entry of one body into the shadow of another. When the Moon is eclipsed, it is passing

through the shadow of the Earth in space. Other planets can also be seen to eclipse their moons. When the Moon moves in front of the Sun this is called an eclipse, although strictly it should be termed an OCCULTATION.
Encke's comet is the comet with the shortest known return period, 3.3 years. It is named after the German astronomer Johann Franz Encke (1791–1865) who calculated its orbit in 1819. Encke's comet moves within the orbit of Mercury at its PERIHELION, but recedes 75 per cent of the dis-

about a quarter the diameter of the Earth. The two bodies have quite an effect on each other. The force of the Earth's gravity has slowed the rotation of the Moon until it rotates once in exactly the same time as it revolves around the Earth. As a result, it always presents the same face to the Earth and the far side is never seen. The Moon's gravity also pulls at the Earth, raising the level of the ocean to produce a high tide beneath the Moon. Another high tide occurs on the opposite side of the Earth at the same time. This is produced by the centrifugal force of the Earth's motion around the centre of gravity of the Earth–Moon system, a point that in fact lies beneath the Earth's surface. As the Earth rotates beneath these two rises in level, most coasts experience two high tides a day. The Sun also affects the tides, raising the level when it is in line with the Moon and Earth.

Below: Moon craters.
A. Most of the many craters on the Moon formed by the impact of meteorites. As a meteorite struck the surface (1), it gouged out a deep crater and threw up boulders as it broke the underlying rock. The boulders then fell back to the lunar surface (2), producing several small craters around the first one.

B. Some of the Moon's craters formed by volcanic action. Pockets of hot gas from the Moon's interior forced up the surface (1), which then collapsed from beneath (2) to form a crater (3).

The Moon and Sun also interact to produce ECLIPSES. An eclipse of the Sun occurs when the Moon passes in front of the Sun. The Moon appears to be about the same size as the Sun from the Earth, and so its shadow or UMBRA only just touches the surface, making a solar eclipse a rare sight. But where it does, a total eclipse occurs: the Sun's disk is blotted out and its CORONA (*see page 24*) becomes visible. On either side of the shadow, in the PENUMBRA, a partial eclipse is seen as part of the Sun's disk is hidden. An eclipse of the Moon is much more likely to be seen, for it can be viewed from anywhere that the Moon is visible, as it dives into the Earth's shadow. The part of the Moon visible to Earth varies in shape because we see changing portions illuminated by the Sun as the Moon goes round the Earth. These changes in shape are called phases, and a complete cycle from new Moon to full Moon and back again takes 29.5 days.

Man has long dreamed of going to the Moon; a story of a Moon journey was written as long ago as AD 150. Little was known about the Moon before the invention of the telescope, and then more than three centuries of subsequent observation was overthrown when orbiting spacecraft mapped the whole of the Moon in the mid-1960s. Preceded by automatic space probes that surveyed the Moon's surface, the first astronauts landed in 1969. Although encumbered by heavy spacesuits, they found it easy to move due to the low gravity – one-sixth of that on Earth. The landscape was striking. Beneath the jet-black sky (the Moon has no atmosphere), the virtually colourless ground was strewn with craters and rocks and covered in a fine dust – all the result of constant bombardment by meteorites over millions of years. The Sun heated the ground by day to more than 100°C, though it would plunge to –150°C during the long night. The lunar day is long because the Moon spins slowly.

A

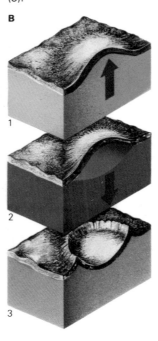

B

Right: The tides vary in extent with the positions of the Sun and Moon. When they are in line with the Earth, their pulls of gravity combine to produce a large rise and fall, giving the spring tides. In-between, the Sun and Moon are at right-angles and their pulls of gravity tend to cancel out. The rise and fall is low, giving the neap tides.

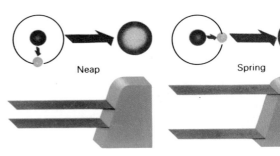

Moon · Sun · Earth · Neap · Spring · Neap · Spring · High tide · Low tide

Earth

Mars

Phobos

Deimos

Above: The Earth and Mars with its 2 moons to scale.

Right: A pair of pictures of the surface of Mars taken by the American space probe Viking 1, which landed on Mars in 1976. The 2 pictures can be combined to give a stereoscopic view of the surface. They show a red desert littered with rocks. The space probe has also observed sand dunes blown by the wind, and falls of snow during the Martian winter.

Right: A view of Mars made by fitting together 1,500 computer-corrected television pictures from the American space probe Mariner 9, which orbited Mars from 1971 onwards. The north polar cap is at the top.

Bottom right: This picture of the surface of Mars was taken by the space probe Viking 1. It shows the most prominent feature on Mars, the Tharsis mountains and Olympus Mons. The mountains consist of a chain of 3 volcanoes each 20 km high, twice the height of Mount Everest. Olympus Mons, to the top of the picture, is another volcano half as high again, and is the largest known mountain in the Solar System.

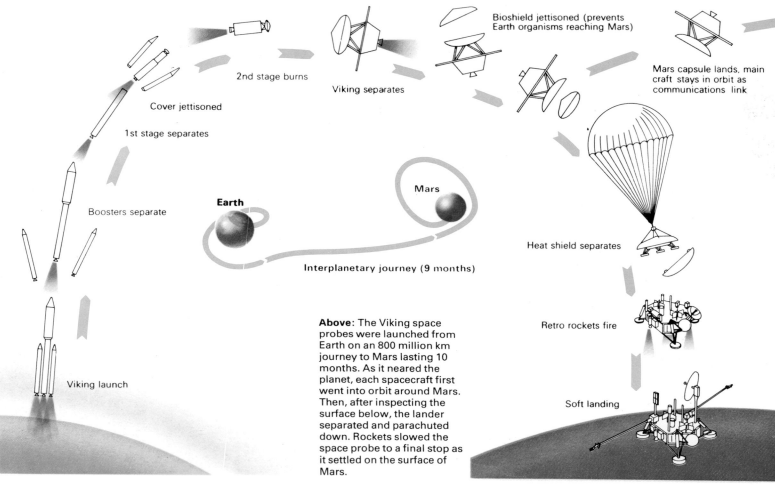

2nd stage burns

Viking separates

Bioshield jettisoned (prevents Earth organisms reaching Mars)

Mars capsule lands, main craft stays in orbit as communications link

Cover jettisoned

1st stage separates

Boosters separate

Earth

Mars

Heat shield separates

Interplanetary journey (9 months)

Retro rockets fire

Viking launch

Soft landing

Above: The Viking space probes were launched from Earth on an 800 million km journey to Mars lasting 10 months. As it neared the planet, each spacecraft first went into orbit around Mars. Then, after inspecting the surface below, the lander separated and parachuted down. Rockets slowed the space probe to a final stop as it settled on the surface of Mars.

---

a satellite's orbit and a planet's equator.

J Jupiter is the largest planet of the Solar System, weighing 2.5 times as much as all the other planets combined. It is 142,800 km in diameter, and has an average density 1.3 times that of water. Jupiter orbits the Sun every 11.9 years at an average distance of 778 km. It is made mostly of hydrogen and helium gas, the same composition as the Sun. Jupiter is the fastest-spinning planet, turning on its axis once every 9 hours

50 minutes at the equator. The planet has 13 known moons, and a possible 14th moon has also been reported.

Four moons of Jupiter

K Kant, Immanuel (1724-1804) was a German philosopher who in 1755 put forward the forerunner of modern theories of the origin of the Solar System. Kant believed the planets had grown from a disk of matter surrounding the Sun, an idea developed by the Marquis de LAPLACE. Kant also believed that the fuzzy nebulae seen in space were separate galaxies like our own Milky Way, an idea now confirmed.

Kuiper, Gerard (1905-73) was an American planetary expert, who in 1948 and 1949, found the moons Miranda and Nereid of Uranus and Neptune respectively. Kuiper discovered the atmosphere of Saturn's

Immanuel Kant

satellite TITAN, realized that the Martian atmosphere was thin carbon dioxide, and showed that Pluto was much smaller than previously im-

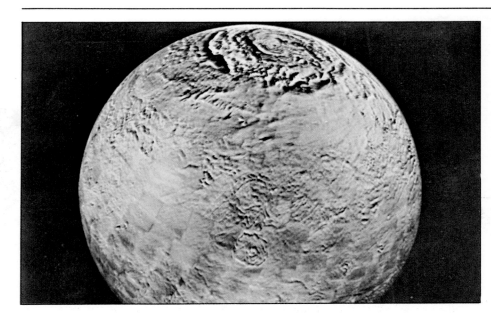

also rotates at almost the same rate as the Earth, so having a day only a little longer than ours, and has two tiny moons, PHOBOS and DEIMOS.

Astronomers have long thought that life might exist on Mars. The average temperature at the surface is –40°C, but it does rise to above freezing point at the equator during the day and some kind of primitive life might be able to survive these conditions. A century ago, many believed that CANALS existed on Mars dug by intelligent beings, but space probes have clearly shown that no such canals are to be found.

Most of our present knowledge about Mars comes from a series of American space probes; none of the Russian missions to Mars have been successful. In 1965, MARINER 4 flew past Mars, sending back the first detailed pictures that showed craters on the surface. More detailed pictures came in 1969, when Mariners 6 and 7

The dozen astronauts who visited the Moon between 1969 and 1972 set up instruments to measure conditions there and brought back samples of Moon rocks to Earth. These missions answered some of our questions about the Moon. The craters are mostly formed by the impact of meteorites, though some seem to be of volcanic origin. The great plains that GALILEO (*see page 7*) thought were seas – and are consequently so named – consist of lava and were formed long ago when huge meteorites struck the surface and broke it open, allowing lava to flow out over large areas. Such activity does not now take place, for the Moon is a dying world. It formed about 4,600 million years ago at the same time as the Earth. Differences in rocks show that it was never part of the Earth. It either formed alongside the Earth or was later captured by the Earth's gravity.

Mars

The planet MARS has long been known as the Red Planet, for its red hue can easily be seen when the planet comes nearest to us at OPPOSITION. It may then come within a range of 56 million kilometres. Telescopes show it to be a small world, half the size of the Earth in diameter. Light and dark reddish markings can be seen, together with two white polar caps. Both the markings and the polar caps vary in extent with the seasons, for the axial tilt of Mars is almost the same as the Earth and similar seasons occur. It

agined, no larger than Mars. Kuiper encouraged the belief that planetary systems were common around many stars and was also an expert on lunar geology.

L Lagrangian points are the gravitationally stable regions where the attractions of the Earth and Moon cancel each other out. The 2 most important points, called L4 and L5 lie 60° ahead and behind the Moon in its orbit. Similar Lagrangian points lie in the orbit of Jupiter, where the TROJAN asteroids collect. The points

are named after the French mathematician Joseph Louis Lagrange (1736-1813) who calculated their existence in 1772.

Laplace, Pierre Simon, Marquis de (1749-1827) was a French mathematician who developed the theory of the origin of the Solar System proposed by Immanuel KANT. In 1796 Laplace described how rings of material thrown off from the Sun might condense into individual planets. The details of the theory have been revised, but the general principle is similar to that of

modern theories of the origin of the Solar System.

Leverrier, Urbain Jean Joseph (1811-77) was a French mathematician who predicted the existence of the planet Pluto. Like the Englishman John ADAMS, Leverrier examined the movement of Uranus, and found that it was being disturbed by the gravitational pull of an unknown planet. Leverrier's calculations allowed Johann GALLE to discover Neptune.

Lowell, Percival (1855-1916) was an American astronomer who drew maps of

Percival Lowell

canals on Mars, and believed in the existence of life on that planet. In 1894 he set up the Lowell Observatory in Arizona to study Mars.

Lowell also believed in the existence of a hitherto undiscovered planet beyond Neptune, and began a photographic search of the sky for it. The new planet, named Pluto, was eventually found after his death by Clyde TOMBAUGH.

M Magnetosphere is the area around the Earth where our planet's magnetic field stretches out into space. The magnetosphere deflects the solar wind of atomic particles that stream outwards through the Solar System from the Sun, and

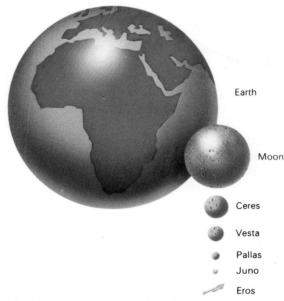

Earth

Moon

Ceres

Vesta

Pallas

Juno

Eros

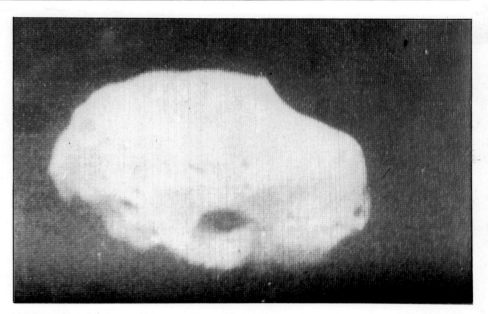

flew past. Then in 1971, Mariner 9 went into orbit around Mars and was able to map the whole of the planet.

Most of the Martian surface is covered with craters, but they are generally not as deep as the craters on the Moon, presumably because they have been eroded by sandstorms. Among the larger features is a vast low plain called Hellas, which was formed by meteorite impact in the same way as the 'seas' on the Moon. On the opposite side of Mars to Hellas are four huge volcanoes, of which the largest, Olympus Mons is 500 kilometres across and three times as high as Mount Everest. The volcanoes may still be active. Between Hellas and the volcanoes is a high plateau cut by a deep rift valley 5,000 kilometres long. From the plateau and elsewhere extend sinuous valleys that resemble water channels. Water may have existed there in the past, but on Mars it is now locked up in the polar caps, which are formed of ice and frozen carbon dioxide. The caps recede and advance as spring and autumn arrive, the ice and carbon dioxide evaporating into the thin atmosphere for the summer but condensing on the ground for the winter. The seasonal colour changes that are visible from Earth are thought to be due to dust storms.

In 1976, two American VIKING probes landed on Mars and sent back the first close-up pictures of the surface. Both showed a red desert littered

Above: A view of Phobos, the larger moon of Mars, taken by Mariner 9. Phobos was discovered by the American astronomer Asaph Hall in 1877. Later pictures show a series of lines on Phobos, which could possibly be cracks and indicate that the moon is about to break up

Above left: The Earth, Moon and 5 asteroids to scale. Ceres, Pallas, Vesta and Juno are the largest asteroids, and Eros is an irregular asteroid that comes near the Earth.

with rocks. They scooped up some soil and scanned it for living organisms, but without positive success. Future missions to Mars planned by the United States include the placing of two automatic vehicles on the surface in 1985, and the return of Martian soil to Earth in 1988. A manned landing might follow in the next century.

The asteroids

Between the orbit of Mars and Jupiter lies a belt of minor planets known as the ASTEROIDS. The largest, CERES, is 1,000 kilometres across. Other large asteroids include Pallas, Vesta and Juno, but few are bigger than 100 kilometres in diameter. Most are great chunks of rock, like mountains floating in space. Most asteroids remain within the asteroid belt, but a few wander outside. One called Icarus takes its name from the Greek legend of Icarus, who flew too near to the Sun. At PERIHELION this asteroid comes nearer the Sun than any other body and may glow red-hot. Some other asteroids wander in the other direction; Hidalgo, for instance almost reaches the orbit of Saturn. Several asteroids cross the Earth's orbit, and Hermes came to within one million kilometres of Earth in 1937. However, the chances of an asteroid colliding with Earth are very, very remote.

BODE'S LAW made people wonder whether the asteroids might be the remains of a planet that

---

traps some of them forming the VAN ALLEN BELTS.
Mariner spacecraft are a series of planetary probes. Mariner 2, the first success of the series, was the first probe to reach another planet, sending back information about Venus in December 1962. Mariner 4 in July 1965 revealed craters on the surface of Mars, which were studied in more detail by Mariners 6, 7, and 9. Mariner 10 in 1974 flew past Venus, returning photographs of its clouds, before reaching Mercury to survey its surface in detail for the

first time. Mariners 11 and 12, sent to the outer planets, were renamed VOYAGER.
Mars is the fourth planet in distance from the Sun, orbit-

'Giant footprint' crater on Mars

ing every 687 days at an average distance of 228 million km. Mars is a small planet, 6,787 km in diameter and rotates once every 24

hours 37 minutes, only slightly slower than the Earth. Mars is a rocky body with a thin atmosphere of nitrogen and carbon dioxide. It has polar caps consisting of frozen water and frozen carbon dioxide. The surface of Mars is covered with craters formed by meteorite impact and by volcanoes, and is commonly known as the red planet, because of the colour of its desert-like surface. Mars has two moons, PHOBOS and DEIMOS. Space probes have failed to detect any signs of life on the planet.

Mercury is the closest planet to the Sun, orbiting every 88 days at an average distance of 57.9 million km. It is very small, only 4,880 km in diameter, and the Sun's gravitational pull has slowed its axial rotation so that it spins only once every 59 days. It has scarcely any atmosphere, and no known satellites. The average density of Mercury is 5.5 times that of water, and probably has a very large core of iron.
Meteor is a tiny particle from space burning up by friction with the Earth's atmosphere. The average

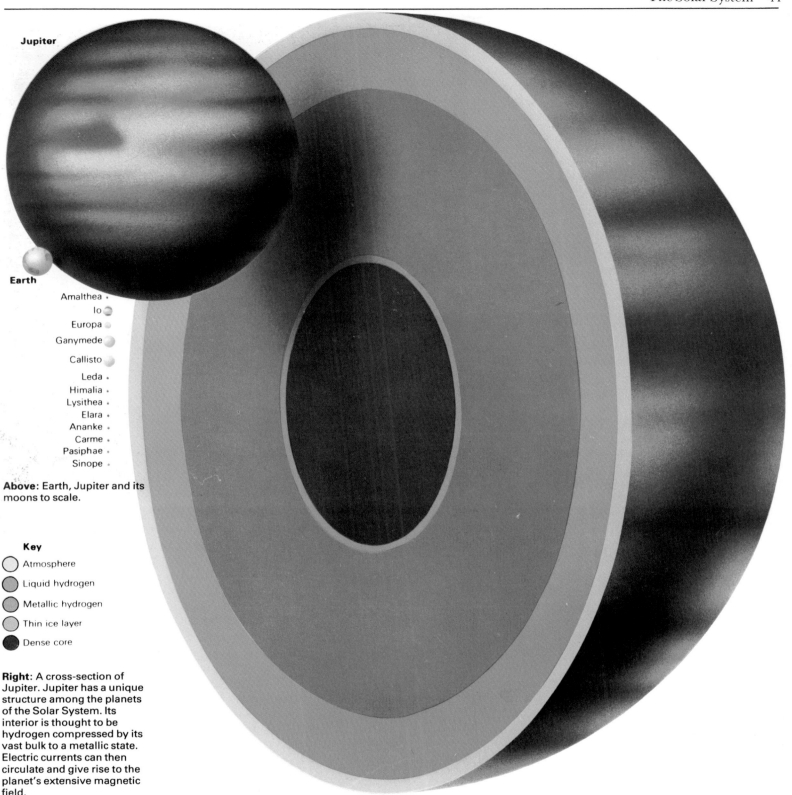

Jupiter

Earth

Amalthea ·
Io ◦
Europa ◦
Ganymede ◦
Callisto ◦
Leda ·
Himalia ·
Lysithea ·
Elara ·
Ananke ·
Carme ·
Pasiphae ·
Sinope ·

Above: Earth, Jupiter and its moons to scale.

Key

◯ Atmosphere
◯ Liquid hydrogen
◯ Metallic hydrogen
◯ Thin ice layer
● Dense core

Right: A cross-section of Jupiter. Jupiter has a unique structure among the planets of the Solar System. Its interior is thought to be hydrogen compressed by its vast bulk to a metallic state. Electric currents can then circulate and give rise to the planet's extensive magnetic field.

meteor is about the size of a grain of sand, and burns up at a height of about 100 km. Meteors are debris from comets, and meteor showers occur when the Earth crosses a comet's path. One of the most famous meteor showers is the Perseids, visible every August at the rate of one every few minutes coming from the direction of the constellation Perseus.
Meteorite is a large lump of rock or metal from space that penetrates the Earth's atmosphere. Most meteorites are believed to be debris from the formation of the

Solar System, but some may be pieces from the heads of comets. They arrive in the Earth's atmosphere at random, and their appearance cannot be predicted like that of meteor showers, which are caused by much smaller bodies. The heaviest known meteorite to have landed on Earth lies at Hoba West near Grootfontein in Namibia (South-West Africa); it weighs 60 tonnes and was moving too slowly to form a CRATER on impact.
Month is a division of time based on the Moon's cycle of phases, which take 29

days to go from new through crescent to full and back to new again. This is the 'synodic month'. The 'sidereal month' is the time taken for the Moon to return to the same position against the star background, and lasts 27.3 days. Early calendars were based on the synodic month. The modern calendar month is an artificial device designed to fit a year of 365 days.
Moon is the natural satellite of the Earth. It is 3,476 km in diameter, and orbits the Earth at an average distance of 384,400 km every 27.3

days. The Moon also spins on its own axis in this time, so that it keeps one face permanently turned towards the Earth. The Moon prob-

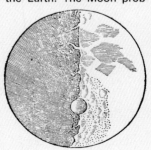

Galileo's drawing of the Moon

ably formed in orbit around the Earth, although it may have been an independent body that was subsequently captured by the Earth's gravity.

N **Neptune** is the eighth planet in distance from the Sun. It is a giant planet made of gas, 49,500 km in diameter. It consists mostly of hydrogen and helium, although there is probably also methane in its outer atmosphere. Neptune probably has a rocky core, is covered in ice, and orbits the Sun every 164.8 years at an

long ago broke into fragments, but their total mass is much too small. They could be fragments of rock that never formed into planets because of the disturbing gravitational influence of Jupiter. However, it is possible that there were once several asteroids the size of Ceres which later collided and broke apart.

Jupiter

The planet JUPITER is the first of the outer planets of the Solar System, which are different in many ways from the inner planets. They are mostly very big, Jupiter being the largest of all the planets with a diameter 11 times that of the Earth and a volume over 1,000 times greater. Around it orbit 13 or possibly 14 moons, of which the four largest were discovered by Galileo and named Io, Europa, Ganymede and Callisto. Jupiter also spins more rapidly than any other planet, making it bulge at the equator.

Through a telescope, Jupiter can be seen to be crossed with bands of light yellows, oranges and greys. The bands consist of clouds floating in Jupiter's atmosphere. In the Southern Hemisphere is a region known as the RED SPOT, which is believed to mark the site of an immense permanent storm in Jupiter's atmosphere. The atmosphere consists of hydrogen, helium, water vapour, methane and ammonia, and its temperature is −150°C at cloud level. The surface is hidden beneath the clouds, but it is thought to be composed of liquid hydrogen. In fact, the whole planet is probably composed mainly of hydrogen, which could be compressed into a metallic state in Jupiter's core. However, at the very centre of the planet there could be a core of rock, with a mass about forty times that of the Earth. It is possible that all the inner planets would have a similar structure to Jupiter if they were further away from the Sun. The Sun's heat could have driven away the hydrogen and other light gases, leaving the rocky cores exposed and forming the planets we see today.

Jupiter is also very unusual because it emits radio waves. This shows that Jupiter has intense belts of radiation, which are produced by charged particles trapped in the planet's strong magnetic field. In fact, the amount of energy that Jupiter gives out is much greater than the energy it receives from the Sun. Clearly, it is making its own energy, perhaps from radioactive processes

Above: A picture of Jupiter taken by the American space probe Pioneer 10 as it approached the giant planet. The red spot and the bands of cloud floating in Jupiter's atmosphere are clearly shown.

in its interior or possibly because, being liquid, it is contracting under its own gravity. A contraction in size of only one millimetre a year would produce enough energy to cause this process. The resulting heat gives rise to the weather systems on Jupiter, and it could make the surface warm by our standards.

In 1973 and 1974, the American space probes Pioneer 10 and Pioneer 11 flew past Jupiter sending back pictures and information to Earth. The pictures showed details of the cloud belts and instruments measured conditions such as temperature and magnetism, helping astronomers to formulate better ideas about Jupiter. American VOYAGER space probes now on their way to Jupiter will fly past in 1979 en route for the other outer planets. The United States plans to send a spacecraft to orbit Jupiter in 1984 and send down entry probes into the atmosphere. A

---

average distance of nearly 4,500 million km. Its period of spin is not precisely known; latest figures suggest it rotates once about every 23 hours. The planet has two known moons: Triton, the largest, and Nereid, 500 km in diameter. Because of the eccentric orbit of the planet PLUTO, Neptune will be the outermost planet of the Solar System during the period of 1979 to 1999.
Nutation is a nodding of the Earth's axis in space, caused by the gravitational pull of the Moon.

Earth from space

O **Occultation** is the obscuration of one celestial body as another passes in front of it. The Moon regularly occults stars as it moves around the Earth. An eclipse of the Sun by the Moon is an occultation.
Opposition is the moment when a planet appears opposite the Sun in the sky. At opposition, a planet appears due south in the sky at midnight from the Northern Hemisphere (or due north from the Southern Hemisphere). Only objects farther from the Sun than the Earth can come to opposition.

Orbit is the path of one body around another under the influence of gravity. All orbits in the Solar System are ellipses. Some comets have highly elliptical orbits which may, at their farthest points, take them halfway to the nearest stars.

P **Penumbra** is the light outer part of a shadow. When the Moon casts its shadow on the Earth during a solar eclipse, places within the penumbra see only a partial eclipse.
Perigee is the closest point that an object comes in its

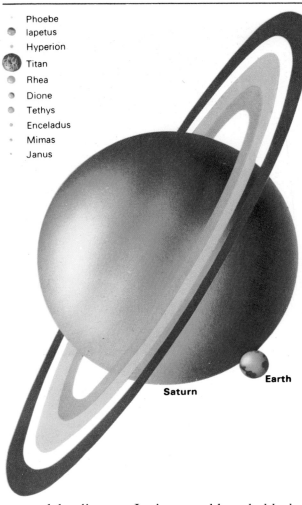

Phoebe
Iapetus
Hyperion
Titan
Rhea
Dione
Tethys
Enceladus
Mimas
Janus

Saturn   Earth

Left: The Earth, Saturn and its moons to scale.

Below: A photograph of Saturn taken through a telescope shows the planet and its rings in some detail. The first space probe will arrive at Saturn in 1979 and should send back close-up views.

The rings are not solid, but made up of millions of rock fragments probably coated with ice. The divisions between them are caused by the gravitational influence of Saturn and its moons. The axial tilt of Saturn is 27° and its period of revolution 29.5 years, so during this period we see first the topside and then the underside of the rings tilted at this angle. In between, the angle is less and every 15 years, we see the rings edge on. Being no more than ten kilometres thick, they become almost invisible at this time. The rings could be composed of the remains of a moon that broke up because it came too near to Saturn and passed ROCHE'S LIMIT. Alternatively, it could be composed of fragments from the cloud that formed the planets and moons; the proximity of Saturn produced a ring system instead of another moon.

Saturn has ten moons in all. The largest, TITAN, is unusual because it has an atmosphere of methane and possibly hydrogen. It has a ruddy colour and clouds may even float in the atmosphere. Titan is too small to retain such an atmosphere, so the gases must be continually produced by the planet. In addition, Saturn's gravity may keep the gases in Titan's orbit. Iapetus, another moon, is also interesting because one side appears to be brighter than the other. Either the two sides are different colours or they have different shapes. Phoebe, the outermost moon, is probably a captured asteroid.

manned landing on Jupiter would probably be impossible, but astronauts might one day visit the nearer moons.

Saturn

The planet SATURN is the most impressive sight in our Solar System with its magnificent rings. Saturn's existence has been known since ancient times, but the rings cannot be seen with the naked eye. Galileo was the first to spot them, but his telescope did not give a detailed view and he thought they were extra bodies attached to Saturn, forming a 'triple planet'. The first astronomer to make out the ring system was Christiaan Huygens in 1655. Jean CASSINI found that there is one ring inside another, separated by a gap, now called Cassini's division. In fact, two fainter rings have been discovered nearer to Saturn, making four rings in all.

---

orbit around the Earth. It is opposite to the APOGEE.
Perihelion is the closest point that an object approaches in its orbit around the Sun. It is opposite the APHELION.
Phobos is the larger and nearer of the 2 moons of Mars, discovered in 1877 by Asaph Hall. Phobos orbits Mars every 7 hours 40 minutes, less than 6,000 km above the planet's surface. Phobos is an irregular-shaped body approximately 23 km across, covered with craters. It was discovered in 1877 by Asaph Hall.

Pioneer spacecraft are a series of American probes to explore the Solar System. The first 4 Pioneers, launched in 1958 and 1959, were Moon probes and all were failures. Pioneers 5 to 9, launched between 1960 and 1968, were interplanetary probes, monitoring solar activity from between the planets. Pioneers 10 and 11, launched in March and April 1973 and 1974 respectively, produced our first close-up look at the giant planet Jupiter. Pioneer 10 continued into outer space, while Pioneer 11 flew to Saturn.

Planet is a dark body that shines only by reflecting starlight. Planets are believed to form as natural by-products of the formation of

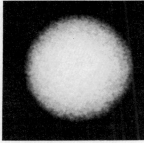

Uranus

stars, so that many stars may have planetary systems. There are 9 major planets orbiting our Sun; together with the other debris and satellites, they are termed the SOLAR SYSTEM.
Pluto is the planet with the greatest average distance from the Sun, 5,900 million km. However, its orbit is the most eccentric of any planet, and the closest it can come to the Sun is 4,400 million km. It is so distant that it takes 248 years to orbit the Sun. The planet rotates once every 6.4 days, but its diameter is not known with

certainty; latest measurements suggest it is only about 3,000 km across, which would make it the smallest planet of all. Pluto was discovered in 1930 by Clyde TOMBAUGH. Many astronomers believe it may be an escaped satellite of Neptune.
Precession is an effect produced by the combined gravitational pulls of the Sun and Moon that make the Earth swing round on its axis in space like a spinning top. Because of precession, the positions of the celestial poles are slowly but continu-

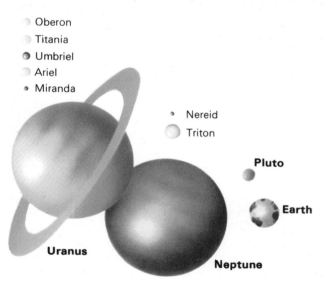

Oberon
Titania
Umbriel
Ariel
Miranda

Nereid
Triton

Pluto

Earth

Uranus

Neptune

Left: The Earth with Uranus, Neptune and Pluto to scale.

Right: An artist's impression of Uranus seen from the ring that surrounds it. The ring is very faint and was only discovered in 1977. It probably consists of fragments of rock floating in space around the planet, like the rings of Saturn.

ally changing. As the Earth swings round in space, the position at which the Sun's path crosses the celestial equator (the EQUINOXES) is also changing; this is why the effect is sometimes referred to as the precession of the equinoxes.

R **Red spot** is a prominent, oval-shaped feature in the clouds of Jupiter, several times the size of the Earth. The spot was first seen in 1666 by Jean CASSINI and is apparently a permanent feature. The spot is several kilometres higher

than the surrounding clouds of Jupiter. It is thought to be a storm cloud caused by an updraught of warm air, as in a terrestrial thunderstorm.

Jupiter's red spot

The spot's colour may be caused by red phosphorus.

Retrograde means axial rotation or orbital motion from east to west, opposite to the standard west-to-east direction of the Solar System. For example, Venus has a retrograde spin, and some of the satellites of the outer planets orbit in retrograde fashion. The term is also used to describe the apparent backwards motion of planets farther from the Sun as the Earth overtakes them on its orbit.

Roche's limit is the distance from a planet at which

a moon would be shattered by tidal forces. For 2 bodies of similar composition, Roche's limit is 2.5 times the planet's radius. Artificial satellites are not torn apart inside Roche's limit because they are small and strong enough to hold together. Roche's limit is named after the French astronomer Edouard Roche (1820–83) who in 1848 calculated its existence.

S **Sagan,** Carl (1934–) is an American astronomer who has strongly advocated the possibility of

life elsewhere in the Universe. Sagan is a planetary expert who in 1960 predicted that the atmosphere of Venus trapped heat to make

Carl Sagan

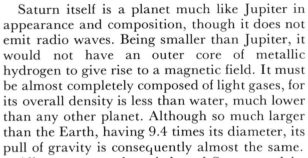

Left: An artist's impression of the surface of Pluto, the most distant planet from the Sun. The Sun is a tiny bright globe in the heavens among the myriad stars of the Milky Way. It is so cold that the ground is covered with frozen gases.

Right: An impression of the planet Neptune as seen from Triton, its largest moon. Little more than its blue colour can be seen from Earth, even through the most powerful telescopes. Our first close-up views could come in 1989 when Voyager 2 should fly past the planet.

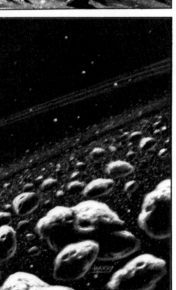

Saturn itself is a planet much like Jupiter in appearance and composition, though it does not emit radio waves. Being smaller than Jupiter, it would not have an outer core of metallic hydrogen to give rise to a magnetic field. It must be almost completely composed of light gases, for its overall density is less than water, much lower than any other planet. Although so much larger than the Earth, having 9.4 times its diameter, its pull of gravity is consequently almost the same.

All our present knowledge of Saturn and its rings and moons is based on telescopic and radar studies. This should change in 1979 when the American space probe Pioneer 11 flies past Saturn five years after its encounter with Jupiter. Voyager 1 should follow it in 1980, having received such a boost in speed as it passes Jupiter that it will make the Jupiter to Saturn journey in only a year and a half.

Uranus

The planet URANUS can just be seen by the naked eye but was not recognized as a planet until it was discovered by telescope by Sir William HERSCHEL (*see page 8*) in 1781 whilst mapping the constellation of Gemini. It is probably similar in composition to Saturn and it also has a faint ring system, discovered in 1977. Uranus also has five moons, Miranda, Ariel, Umbriel, Titania and Oberon, each smaller than our Moon, which revolve round the plane of the planet's equator.

The most unusual feature of Uranus is its axial tilt of 98°. This means that its axis is keeled right over until its north pole points below its orbital plane. The rings and moons therefore lie in a plane almost at right angles to its orbit. The surface of the planet at each pole therefore has a night lasting 21 Earth-years. The greatest tilt of any other planet is Pluto's 50°. No explanation of the tilt of Uranus has been found. The American space probe Voyager 2 may reveal more about Uranus. Now on its way to Jupiter, it should fly past the giant planet and be given a boost towards Uranus, reaching it in 1986.

Neptune

The orbit of Uranus turned out to be erratic, suggesting to astronomers that it was being influenced by the gravitational pull of another unknown planet. John ADAMS and Urbain LEVERRIER separately calculated where such a planet should be found, and it was discovered near the predicted position by Johann GALLE in 1846. The new planet was called NEPTUNE. No telescope is able to pick out definite surface details. All that can be seen is a pale bluish disk. It is probably very similar to Uranus and has two moons, of which TRITON is probably the biggest moon in the Solar System. Nereid is the smaller moon and has an eccentric orbit like that of a comet. VOYAGER 2 may fly on to Neptune after passing Uranus, reaching it in 1989.

---

the planet's surface roastingly hot. He has also studied the formation of complex molecules in Jupiter's atmosphere, which give the clouds their varied colours.
Saturn is the sixth planet in distance from the Sun, and the outermost one known to the ancients. Saturn orbits the Sun every 29.5 years at an average distance of 1,430 million km. Saturn is made mostly of hydrogen and helium gas, similar to Jupiter, although it is somewhat smaller. Saturn's diameter is 119,300 km, and it rotates

once every 10 hours 14 minutes at the equator. Saturn's most famous feature is its rings, 275,000 km from rim to rim. Saturn has 10 known moons, the largest being TITAN.
Schiaparelli, Giovanni (1835–1910) was an Italian astronomer who first reported the "canals" of Mars at the close approach of the planet in 1877. He called them *canali* which in Italian means channels; he did not believe them to be artificial, but the mistranslation gave the erroneous impression. Schiaparelli also showed

that meteor showers follow the same orbits as comets, and proposed that they are debris from comets.
Sidereal period is the time taken for an object to com-

Meteorite

plete one orbit relative to the star background. The figures quoted for the orbital period (year) of a planet or of a satellite around a planet are the sidereal periods.
Solar System is the collection of planets, their satellites and other debris, orbiting the Sun. Anything captured by the Sun's gravity is part of the Solar System. An object can lie halfway to the nearest star and still be part of the Solar System.
Solstice is the time of year when the overhead Sun reaches its northernmost or southernmost point. The

summer solstice in the Northern Hemisphere is on June 21, the winter solstice is on December 21.
Synodic period is the time taken for a Solar System body to return to the same position, relative to the Sun, as seen from the Earth. For instance, the synodic period of the Moon is the time for it to undergo one complete cycle of phases. The synodic period differs from the SIDEREAL PERIOD.

T **Tombaugh,** Clyde(1906–) is an American astronomer who in February

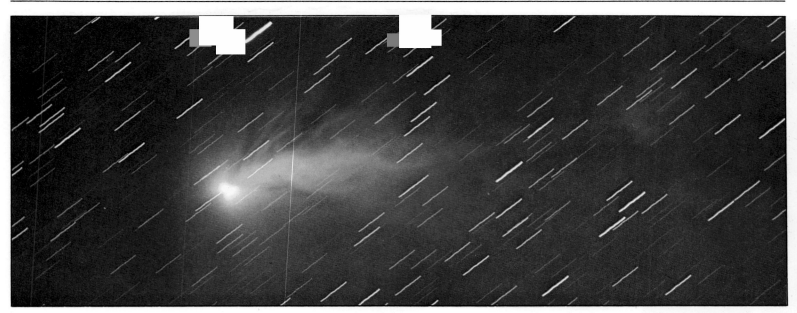

Pluto

The orbit of Neptune was found not to be regular either and yet another unknown planet was therefore predicted. After a long search begun by Percival LOWELL, it was eventually detected by Clyde TOMBAUGH in 1930. The new planet was called PLUTO, and it was found in the predicted position. The search took so long because Pluto was thought to be a large body, but turned out to be very small. Its orbit is very eccentric, taking it inside the orbit of Neptune. In fact, it is thought to have once been a moon of Neptune. Its peculiar orbit has given rise to some doubt as to whether Pluto should be considered a true planet.

The discovery of Pluto leaves an element of mystery. How did it escape from Neptune? And is there another planet beyond Pluto that is responsible for the variations in Neptune's orbit? No space probes to Pluto are planned and no search is likely to be mounted for another planet. The questions may remain unanswered for a long time to come.

Comets

Until quite recently, COMETS were regarded as portents of evil; they had a way of turning up just before disasters occurred. This was coincidence, of course, but it was not until two centuries ago that the appearances of comets were found to be regular events. Edmund HALLEY realized that one

Above: A photograph of Humason comet, which appeared in 1961. A long exposure was necessary to take the photograph because the comet was dim, and during this time the comet moved against the background of stars. The telescope followed the comet, thus making the stars appear to move and produce trails of light.

Below: As a comet moves around the Sun, its tail streams out away from the Sun.

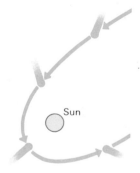

comet had been seen every 76 years for several centuries. He believed that it was the same body returning every 76 years and predicted its reappearance in 1758. Unfortunately, he did not live long enough to see his prediction come true, but the comet in question is named HALLEY'S COMET in his honour. It will next be seen in 1986. Many other comets have been recognized since Halley's discovery. The most frequent is ENCKE'S COMET, but it is invisible to the naked eye.

A comet moves around the Sun in a highly elliptical orbit. It spends most of the time far away from Sun, many comets receding beyond the orbit of Pluto. According to the standard theory of comets proposed by Fred WHIPPLE, it consists of a 'dirty snowball' of dust embedded in ice and frozen gases. As this ball nears the Sun, the ice vaporizes and the gases are released to form a cloud. The pressure of the solar wind makes a trail of dust and gas stream back from the head of the comet. The dust is lit up by the Sun, the gases begin to glow with energy and the comet becomes visible in the heavens. It does not dash across the sky, as its appearance suggests, but moves slowly against the backdrop of stars, remaining in view for a few weeks if it is very bright.

As the comet passes the Sun and recedes, it fades and condenses to a dirty snowball again. The solar wind blows away some of its material so that recurring comets get fainter with each

---

1930 discovered the planet Pluto using photographs taken at the Lowell Observatory. After the discovery of Pluto, Tombaugh continued

Clyde Tombaugh

his photographic survey around the entire sky in search of other possible undiscovered planets, but found none. As a result of his work, most astronomers are confident that no major Solar System bodies remain undiscovered.
Titan is the largest satellite of SATURN, and the only moon known with certainty to have a dense atmosphere. Titan is 5,800 km in diameter, and orbits Saturn every 15.95 days at an average distance of 1.2 million km. In 1944 Gerard KUIPER found that Titan has a substantial

atmosphere of methane; hydrogen may also be present. With the exception of TRITON, Titan may be the largest satellite in the Solar System.
Transit is the crossing of a satellite in front of a planet, or the movement of a planet across the face of the Sun. The term can also be used to mean the moment when a star lies exactly on the meridian (north-south line).
Triton is the largest satellite of NEPTUNE, and possibly the largest satellite in the Solar System. Modern measurements suggest it has a

diameter of 6,000 km. Triton orbits Neptune every 5.88 days in a RETROGRADE (east to west) direction. Triton lies 355,000 km from Neptune.

Trojans are a family of asteroids moving in the same orbit as Jupiter at the LAGRANGIAN POINTS. They are named after heroes of the

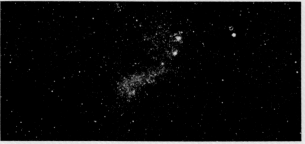

Lesser Magellanic Cloud

return. However, some may have a stony core and evolve into asteroid-like bodies crossing the orbits of the planets.

Comets are frequent visitors to the vicinity of the Earth, but few are bright enough to be seen. The last easily visible comet was in 1957, and Halley's comet could well be the next one. The United States plans to send a space probe to fly past Halley's comet and examine it from close range.

Meteoroids

Meteoroids are pieces of rock, ranging in size from grains of sand to huge boulders, that move through space. The smaller fragments burn up as they encounter the Earth and streak through its atmosphere, producing a 'shooting star' or METEOR. Larger bodies that survive their fiery flight and strike the ground are called METEORITES. Meteors are a frequent sight; usually, at least one occurs in an hour of sky-watching. They are rapidly moving points of light, often with luminous tails. Meteorites are rare, but may cause severe damage. They are not merely large meteors.

Meteors are rock particles from comets. After comets pass around the Sun, clouds of particles may be left behind. When the Earth passes through such a cloud, a meteor swarm occurs and a host of meteors appears from a certain direction in the heavens. The swarms are regular and are named after the constellations in which they appear to originate; for example, the Leonid swarm appears to come from the constellation Leo. This swarm can be seen every 33 years and has been spectacular in the past. Although 1899 and 1933 were poor years, 1966 unexpectedly produced a rich harvest of meteors. As many as 40 meteors per second were seen in the United States at the peak of the swarm.

Meteorites are more interesting to astronomers because they probably consist of debris left over when the Solar System formed. They are made of rock or metal, or a mixture of both. The sacred stone at Mecca is almost certainly a meteorite. From the ages of meteorites, scientists believe that the planets and moons formed about 4,600 million years ago within a period of 100 million years.

Where large meteorites strike the ground, they produce huge craters. A huge crater more than

Above: A crater produced by the impact of a meteorite long ago in central Australia. The crater resembles many of those on the Moon, which were produced in the same way.

Below: The path of a meteor, or 'shooting star', across the sky. This photograph shows it exploding as it descends through the atmosphere.

Trojan wars (e.g. Achilles). Only 15 Trojans are named, but there may be several hundred in all.

U Umbra is the dark central part of a shadow. From a spot within the umbra of the Moon's shadow one could see a total solar eclipse. The term is also used to denote the dark central portion of a sunspot.
Uranus is the seventh planet in distance from the Sun. It was found accidentally on March 13, 1781 by William HERSCHEL. Uranus orbits the Sun every 84 years

William Herschel

at an average distance of 2,900 million km. It is a giant ball of gas, 51,800 km in diameter. The spin of Uranus is not known accurately; recent figures suggest it turns on its axis once every 22 hours. Uranus looks slightly greenish through a telescope, so its atmosphere probably contains considerable methane, hydrogen and helium. Underneath its atmosphere, Uranus may have a rocky core, coated with ice. The planet has 5 known satellites, and in 1977 it was found to have thin rings.

V Van Allen Belts are the areas around the Earth where the Earth's magnetic field traps atomic particles from the Sun. The belts are centred at heights of 3,000 and 18,000 km above the Earth. They are named after the American physicist James Van Allen (1914–) who in 1958 discovered them using instruments aboard early satellites.
Venus is the second nearest planet to the Sun, familiar as the brilliant 'morning star' or 'evening star' as seen from Earth. At its brightest it can outshine all other objects

except the Sun and Moon. It orbits the Sun every 225 days at an average distance of 108 million km, and can come closer to the Earth than any other planet. Venus appears so bright because it is swathed in an unbroken blanket of white clouds, believed to be made of sulphuric acid droplets. Beneath its atmosphere of carbon dioxide gas, Venus is a rocky planet 12,100 km in diameter, almost equal in size to the Earth (it has often been termed 'Earth's twin'). Radar shows that the surface is pocked with craters, and

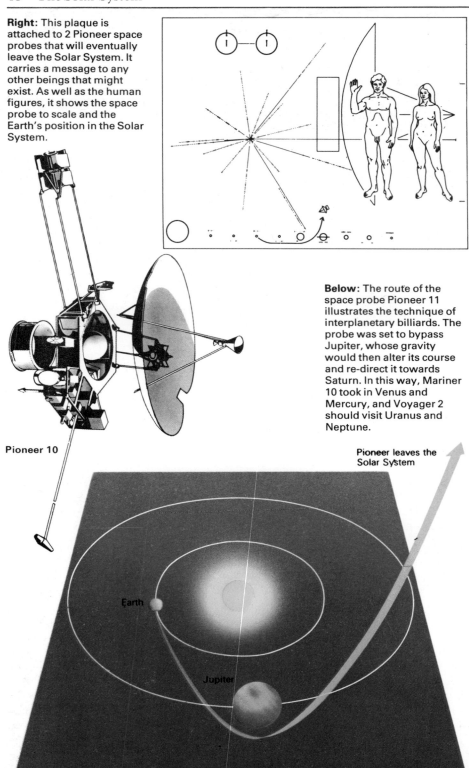

Right: This plaque is attached to 2 Pioneer space probes that will eventually leave the Solar System. It carries a message to any other beings that might exist. As well as the human figures, it shows the space probe to scale and the Earth's position in the Solar System.

Pioneer 10

Below: The route of the space probe Pioneer 11 illustrates the technique of interplanetary billiards. The probe was set to bypass Jupiter, whose gravity would then alter its course and re-direct it towards Saturn. In this way, Mariner 10 took in Venus and Mercury, and Voyager 2 should visit Uranus and Neptune.

Pioneer leaves the Solar System

Earth

Jupiter

400 kilometres across is believed to exist at Hudson Bay in Canada, but the largest proven meteorite crater is the Arizona crater, which is more than a kilometre wide. If the kind of meteorites that hit the Moon in the past and formed its seas and large craters were to strike the Earth now, they would cause tidal waves as high as mountains, and probably wipe out human life.

Life in the Universe

The discoveries of space probes have shown that it is very unlikely that life, certainly as we know it, can exist anywhere else in our Solar System other than on Earth. It exists here because the conditions are right. If the Earth were smaller, colder or less massive, then life would take on different forms. In the early stages of the Earth's history, complex chemical compounds may have built up from simple compounds and gradually developed into living things. The kinds of compounds involved have been found in meteorites, and radio astronomers have detected them in outer space. They indicate that the requirements for life could be spread throughout our galaxy and probably the Universe.

Carl SAGAN, the American astronomer, is very interested in the question of other life in the Universe, and has estimated that there are most likely to be about a million planets inhabited by intelligent beings in the galaxy. If so, they would lie an average distance of 300 light years apart. Our chances of contacting such a civilization – unless its representatives come to us – are very slim, but efforts have begun. Two PIONEER space probes, destined to leave the Solar System, each carry a plaque giving information about our civilization. The VOYAGER probes carry gramophone records of human speech and photographs.

Any interstellar communication is more likely to take place by radio. A message sent by any civilization in the galaxy could be picked up by our radio telescopes. Astronomers have been listening for any such messages, but the Universe has so far remained silent with regard to intelligent signals. We ourselves have sent out messages, but an immediate reply from the nearest star would take nine years. If a civilization lies closer than 300 light years away, the reply would take 600 years to reach us.

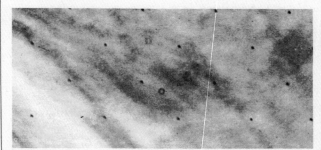

Cloud cover of Venus

also that Venus has a peculiar axial rotation: it spins every 243 days from east to west (RETROGRADE). The planet's dense atmosphere traps heat in what is known as the greenhouse effect, giving it baking hot temperatures of 475°C at the surface. It has no known moons.

Viking probes were 2 spacecraft sent to look for life on Mars. Each Viking was in 2 parts: an orbiter and a lander. While the orbiter section acted as a link to Earth, the lander touched down on Mars, sending back photographs from the surface and carrying out soil analyses. Viking 1 touched down on July 20, 1976, and Viking 2 on September 3. Both probes added considerably to our knowledge of Mars, though neither found evidence of Martian life.
Voyager probes are 2 American planetary probes, launched in 1977 to study the outer planets. Voyager 1 is due to reach Jupiter in 1979 and then fly on to Saturn by 1980. If this is successful, the second Voyager may also be sent on to Uranus and Neptune.

W **Weizsacker,** Carl von (1912–) is a German astronomer who in 1945 proposed the basis of modern theories of the origin of the Solar System. He envisaged that the planets formed by dust particles collecting together from a disk of material surrounding the primitive Sun. This was a modification of the ideas of KANT and LAPLACE.
Whipple, Fred (1906–) is an American astonomer who has proposed the standard theory of comets.

Y **Year** is the time taken for the Earth to orbit the Sun. It takes approximately 365.25 days. By analogy, the orbital periods of other planets are also referred to as their year.

Science fiction became 'science fact' when the USSR launched Sputnik 1 into orbit around the Earth on October 4, 1957. The Space Age has been marked by tremendous achievements, which represent man's first steps towards the stars.

Man in Space

Man has long dreamed of travelling in space, and writers envisaged all sorts of ways in which this might be done. In AD 150 the Greek satirist, Lucian, had a whirlwind take his fictional heroes to the Moon. Bishop Godwin, an English ecclesiastic, used wild geese to tow people there in a book published in 1638. In 1865 Jules Verne fired his astronauts from a huge gun in one of his stories, while H. G. Wells in 1901 used an anti-gravity material, the most fantastic and perhaps prophetic solution of all. In fact, the actual method by which spaceflight can be achieved – by rocket – did not occur to anyone until the turn of the century, even though rockets and the principle by which they work had been known for a long time.

The Chinese invented rockets as early as 1100, following their discovery of gunpowder. They used them against Mongol invaders in a battle in 1232, and rockets were known in Europe by 1300. They were used mainly as fireworks until the early 19th century, when Sir William Congreve, a British artillerist, developed rocket weapons for British forces. Rocket bombardments were successful, particularly against Napoleon's forces, but developments in gunnery ousted the rocket in the later part of the century. Today, the military rocket – now called a guided missile – threatens us all.

Rockets work by the principle of action and reaction stated by Sir Isaac NEWTON (*see page 10*) as one of his laws of motion in 1687. The action of a force in one direction always produces an equal force – the reaction – acting in the opposite direction. A rowing boat moves forwards because the action of the oars pushes water backwards. Jet engines push air backwards to make aircraft move forwards through the air. A vehicle can therefore move through space if it pushes something away from itself. The principle had been shown as long ago as AD 50, when the Greek

Right: The rocket was invented in China in about AD 1200. Mongol invaders besieging a Chinese city in 1232 met 'arrows of fire' that were primitive rockets.

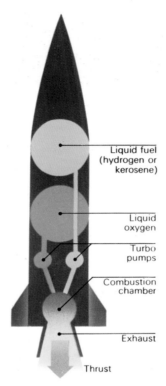

Liquid fuel (hydrogen or kerosene)

Liquid oxygen

Turbo pumps

Combustion chamber

Exhaust

Thrust

Above: A liquid-fuel rocket contains a fuel such as kerosene or liquid hydrogen and an oxidant such as liquid oxygen. The fuel and oxidant are pumped to the combustion chamber where they burn. The hot gases escape through the exhaust, propelling the rocket forward.

engineer Hero built a little engine that worked by spouting jets of steam.

A rocket could also work by boiling water on board and forcing the steam out in a jet from the tail. However, it would not be very powerful. Instead, a rocket burns fuel to produce a jet of hot gases – the hotter they are, the more power they produce. All rockets, from the humblest firework rocket to the greatest spacecraft launcher, work in this way. They need no air to work and so can travel in space. Future rockets may make use of nuclear power to produce a jet of hot gases, or they may have ion engines that contain electric fields to accelerate gases from the tail. One day rockets may be driven by light, as a beam of light consists of a stream of minute particles called photons; this kind of engine is called a photon drive. This could well be the basis for future interstellar travel.

Reference

A Agena is an American upper stage rocket used to launch many satellites, usually on top of an ATLAS booster.

Aldrin, Edwin E. (1930–) was the second man to walk on the Moon. He was lunar module pilot on Apollo 11, which made the first Moon landing in July 1969. Aldrin also made a record 2-hour space walk on the Gemini 12 mission in November 1966.

Ames Research Center is a division of NASA at Moffett Field, California. Ames conducts studies of advanced space projects and researches extraterrestrial life.

Edward Aldrin

The Center is involved in plans for planetary exploration.

Apollo project was the American programme to land a man on the Moon and return him safely to Earth. The United States was committed to this in 1961 by President John F. Kennedy. To accomplish the project, astronauts flew to the Moon aboard the conical 3-man Apollo craft, launched by a SATURN V rocket; 2 men landed on the lunar surface and took off again in the LUNAR MODULE. Apollo 11 made the first manned

Moon landing on July 20, 1969. The last in the series was made by Apollo 17, which landed on the Moon on December 11, 1972.

Ariane is a launch rocket being built by the European Space Agency. The 3-stage rocket is due to begin launching satellites in 1980, from a site at Kourou, French Guiana.

Ariel satellites are a series of British satellites, launched by the United States. Ariel 1, launched on April 26, 1962, was the first satellite of one nation to be launched by another. The most famous

Scorched command module

satellite in the series was Ariel 5, launched in October 15, 1974, for X-ray astronomy studies. Last will be Ariel 6, planned for early 1978.

Tsiolkovsky produced all these ideas as early as 1895, but it was a long time before his work gained attention. Eventually his theories inspired others in Russia, and led to the launching of the first spacecraft.

No one outside Russia knew of Tsiolkovsky's ideas, although some of them did occur to others. In the United States, Robert GODDARD began to experiment with rockets in about 1910. He was not particularly interested in space travel, although he believed a rocket could one day reach the Moon. He saw more immediate use for rockets in exploring the upper layers of the atmosphere and after many years of tests, in 1926 Goddard successfully fired the first liquid-fuel rocket. It rose to a height of 12 metres – not far, but a start; it proved the principle to be valid. By 1935, Goddard's rockets had reached speeds of more than 1,000 kilometres an hour.

Goddard had little backing for his work and could not get any further. However, bodies of

Rocket pioneers

The man who developed the basic ideas behind space travel was a Russian schoolteacher who never fired a rocket in his life – Konstantin TSIOLKOVSKY. He showed that rockets could take man into space, forecasting correctly that they would have to be made up of several smaller rockets or stages mounted on top of each other. As each stage uses up its fuel, which Tsiolkovsky correctly predicted would have to be liquid, it drops away. The rocket is therefore lightened, achieving more power for the same total amount of fuel.

Tsiolkovsky also knew that because the air in a spacecraft cannot be renewed, it would need to be purified, and he forecast that the occupants would be weightless. We feel weight because we are pulled to the ground by the force of gravity and the ground resists any further motion. Except when the engines are firing, a spacecraft and its occupants are equally affected by gravity, whether it be the field of gravity of the Earth, Moon, Sun or a planet. The floor of the cabin moves in the same way as the people inside, and they have no weight because they do not push down on the floor. Unless tethered in some way, they and everything else in the cabin float in mid-air. Astronauts have pre-flight weightlessness training to prepare them for this.

Above: The rockets in a firework display are solid-fuel rockets. They contain gunpowder that burns rapidly when ignited, producing hot gases that drive the rocket upward.

Right: The launch of a V2 rocket from the test site at Peenemünde in Germany during World War II. The V2 was the ancestor of the Saturn rocket that took men to the Moon, as both rockets were designed by the same man, Wernher von Braun.

Right: Syncom 1 was launched by the United States in 1963. These satellites, of which there are several now operating, take 24 hours to make one orbit. They therefore remain above the same spot on the Earth's surface, relaying signals between ground stations on Earth.

rocket enthusiasts had been set up in Germany and Russia and in 1931 and 1933 respectively they fired their first liquid-fuel rockets. Associated with them were two men destined to take the world into the space age – Wernher VON BRAUN and Sergei KOROLEV.

Rocket research now got a helping hand, though not from a welcome direction. In the 1930s the Nazi government took over research in Germany, correctly realizing that rockets had an important role to play in warfare. Von Braun and his colleagues knew that there was no immediate possibility of building rockets for space travel, and during the war period their work produced the V2 rocket. More than 500 of these rockets, armed with high explosive, fell on London in 1944 and 1945, killing thousands of people. At the end of World War II, von Braun went to the United States and there began to develop bigger and better rockets for more scientific purposes, culminating in the Apollo programme.

Meanwhile, the Russians had not been inactive. Korolev worked on rocket-powered aircraft during the war, and headed a team to develop space rockets afterwards. His work had great military value and was given high priority. Unbeknown to the Americans, the Russians were overtaking them.

Below: The rocket that fires a spacecraft into orbit or into space is very large compared with its payload. Early launchers are shown here compared in size to a bus.

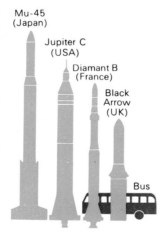

Mu-45 (Japan)

Jupiter C (USA)

Diamant B (France)

Black Arrow (UK)

Bus

Into space

The world finally entered the space age a century after Tsiolkovsky was born. On October 4, 1957, Russia launched the first artificial satellite, Sputnik 1. It was a sphere 58 centimetres in diameter (little bigger than a football) which radioed information on conditions in space back to Earth. Sputnik I stayed in space, orbiting the Earth at a height of up to 1,000 kilometres, for exactly three months.

As early as the 17th century Newton had realized that a satellite could be made to orbit the Earth if it could be launched at a sufficiently great speed. The centrifugal force of its motion would then balance the pull of the Earth's gravity, keeping the satellite up in space. If it stayed high enough, it would be free of the atmosphere and would not slow down and it would remain orbiting the Earth indefinitely. The speed necessary to put a satellite into orbit is 28,200 kilometres per hour and it must reach a height of at least 160 kilometres above the ground. There it completes one orbit every 90 minutes. However, some satellites have elliptical orbits that bring them within the outer reaches of the atmosphere. The drag of the thin air slows the satellite down, and eventually it re-enters the atmosphere and burns up unless it is designed to be recovered.

Cernan, Eugene A. (1934–) is an American astronaut who in December 1972 commanded the final Apollo mission to the Moon, Apollo 17. Cernan also flew in space on the Gemini 9 mission in June 1966 and the Apollo 10 Moon-landing rehearsal in May 1969.
Collins, Michael (1930–) is an American astronaut who was command module pilot on the Apollo 11 mission. He was the one member of the 3-man crew who did not land on the Moon, but instead remained orbiting above. Collins first

flew in space on Gemini 10 in July 1966. He is now director of the Smithsonian Air and Space Museum in Washington, DC, USA.

Apollo command module

Command module was the conical crew compartment of the Apollo spacecraft, 3.9 metres high and 3.2 metres across. To the command module was attached a cylindrical service module, which provided power during the mission and housed the main engine for course changes. The service module was jettisoned before re-entry to expose the command module's heat shield. It was the only part of the Apollo spacecraft to return to Earth.
Conrad, Charles (1930–) is an American astronaut

Charles Conrad

who commanded the second Moon-landing mission, Apollo 12, in November 1969. Conrad also flew on Gemini 5 and

Gemini 11, and commanded the first crew which spent a month in the Skylab space station in 1973.
Cooper, Leroy Gordon (1927–) is an American astronaut who in May 1963 made the longest flight of the MERCURY series, 34 hours. Cooper commanded the Gemini 5 mission in August 1965.
Copernicus is the popular name for the third Orbiting Astronomical Observatory (OAO-3), launched in August 1972.
Cosmos satellites are a continuing series of Soviet Earth

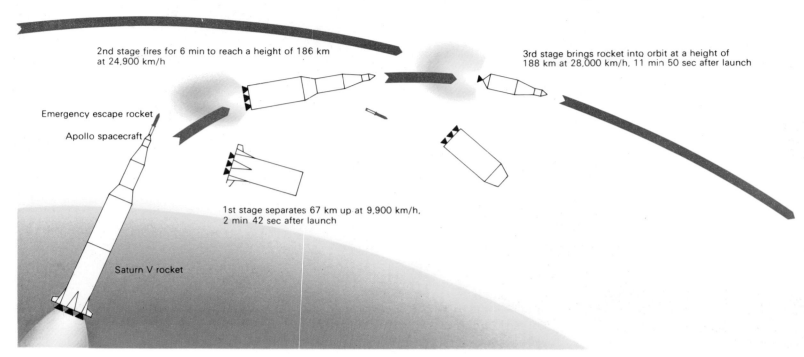

2nd stage fires for 6 min to reach a height of 186 km at 24,900 km/h

3rd stage brings rocket into orbit at a height of 188 km at 28,000 km/h, 11 min 50 sec after launch

Emergency escape rocket

Apollo spacecraft

1st stage separates 67 km up at 9,900 km/h, 2 min 42 sec after launch

Saturn V rocket

The Russians quickly followed up the success of Sputnik 1. Sputnik 2 went into orbit on November 3, 1957 carrying a live dog called Laika. The satellite was large, weighing half a tonne. Instruments measured Laika's reactions to spaceflight, but technology was not yet sufficiently advanced for her to be able to survive the satellite's re-entry.

Meanwhile the Americans were having trouble. Astounded by the Russian success, they pushed ahead with their own satellite programme at full speed but had several mishaps. The first American satellite was launched on January 31, 1958. Called EXPLORER 1, it was a tiny satellite, only 15 centimetres across and weighing only 14 kilograms. However, it made the first major discovery in space: that intense belts of radiation surround the Earth. The belts are named the Van Allen belts after James Van Allen, who headed the research team involved. The next successful American satellite, VAN-GUARD 1, followed Explorer into space on March 17, 1958. It was even smaller than its predecessor, but flew out to nearly 4,000 kilometres from the Earth, providing information on our immediate environment in space. Communication satellites were the next step.

Above: The launch of the Apollo 11 mission that first landed men on the Moon. The Saturn 5 rocket that launched the Apollo spacecraft contained 3 stages. The spacecraft was also attached to an escape rocket that could have fired them away from the Saturn if anything had gone wrong during launch. The 3 stages fired in a precise sequence to place the spacecraft into orbit, and the 3rd stage later fired again to thrust it on its way to the Moon. Each stage and the escape tower were jettisoned when they were no longer needed, so that fuel would not be wasted. They were not recovered.

Earth satellites

These first satellites were the ancestors of a great family of satellites. Since then, well over a thousand have been launched and the family has divided into several branches. Many are scientific satellites. They explore the region of space around the Earth, measuring such factors as the intensity of the Earth's magnetic field, radiation and micrometeoroids. The DISCOVERER satellites were in this class, as were the Explorer and Vanguard. Other scientific satellites look outwards and conduct astronomical studies, especially with radiations, such as X-rays, that do not penetrate the Earth's atmosphere. These satellites include the COPERNICUS, HEAO, OAO, OSO and SMALL ASTRONOMY satellites as well as the projected SPACE TELESCOPE.

The LANDSAT satellites look down to the Earth, seeking useful information that can only be obtained by photographing the ground from space. Weather satellites such as NIMBUS and TIROS also look down, photographing the pattern of clouds. From these pictures, meteorologists now make improved weather forecasts and watch the build up of hurricanes. Military satellites also photograph the Earth, seeking signs of military activities.

Russian space dogs

Yuri Gagarin

Communications satellites provide world-wide television networks. The satellites receive signals from ground stations on one side of the world and transmit them to stations on the other side of the world. Many international telephone calls also go via communications satellites. The INTELSAT satellites lie in GEOSTATIONARY orbits so high that they take exactly 24 hours to go once around the Earth, thus remaining above the same point on the surface and constantly in touch with their ground stations. Early communications satellites included TELSTAR and EARLY BIRD.

Satellites have small rocket motors that enable them to change orbit if necessary and to maintain the alignment required by their guidance systems. Electrical power for the various instruments comes from panels of solar cells that convert sunlight into electricity.

All the satellites named above are American satellites, but the Russians have also launched many kinds of satellite, mostly under the name COSMOS. France, Britain, China and Japan have also made their own way into space.

The space race

With the launchings of the first satellites, rivalry quickly developed between Russia and the United States. Russia maintained its lead in this space race with the three LUNA moon probes of 1959. Following a series of American failures, Luna 1 became the first spacecraft to achieve a speed that enabled it to leave the Earth's gravity (known as ESCAPE VELOCITY). It passed close to the Moon, and was soon followed by Luna 2, which crashed on the Moon's surface. Then came Luna 3, which passed behind the Moon and sent back our first glimpse of the far side.

Russia was able to secure all these firsts because it had developed more powerful rockets than the United States. But American scientists had to construct compact instruments and spacecraft systems to make the best use of their smaller rockets. The sophisticated technology that they were building was to stand them in good stead. The Russians' systems technology was not always up to their fire power. For example, the Luna picture of the Moon was so blurred that it showed few features. However, it was adequate enough to give Russia the next and perhaps greatest prize in space – the first manned spaceflight.

Above: The launch of a Vanguard satellite early in the US space programme. The rocket stands beside a gantry that feeds it with power and fuel.

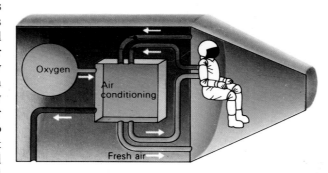

Left: The life support system of a spacecraft removes wastes but recycles the air in the cabin, purifying it to remove carbon dioxide and adding fresh oxygen.

series of US scientific satellites. Explorer 1, launched on January 31, 1958, was the first American satellite to reach orbit. It discovered the Earth's VAN ALLEN BELTS (see page 47). Other satellites in the series have continued to study the Earth and its environment in space. Explorer 55 was launched in November 1975.

G **Gagarin,** Yuri (1934–68) was a Soviet cosmonaut and the first man in space. On April 12, 1961, he completed one orbit around the Earth in VOSTOK 1. Gagarin

Owen Garriot spacewalking

was tragically killed in a plane crash while preparing for another space mission.
Gemini project was the name given to a series of American two-man space missions. In a total of 10 manned flights between March 1965 and November 1966, Gemini astronauts practised the techniques of rendezvous, docking, and working in space, as well as setting up new endurance records for spaceflight. The conical crew section of the Gemini craft was 1.8 metres long and 2.3 metres wide. The Gemini capsules were launched by TITAN rocket.
Geostationary orbit is an orbit 35,900 km above the Earth's equator in which a satellite moves at the same

Gemini rendezvous

rate as the Earth rotates. Therefore the satellite appears to hang stationary above the equator. The geostationary orbit is fre-

quently used for communications satellites. It is also termed a synchronous orbit.
Glenn, John H. (1921–) was the first American to orbit the Earth. He went round it 3 times in the spacecraft Friendship 7 on February 20, 1962. His was the third launch in the MERCURY programme.
Goddard, Robert H. (1882–1945) was an American rocket pioneer. He built and flew the world's first liquid-fuelled rocket, which was launched on March 16, 1926. Goddard built rockets up to 6.6 metres in length.

Left: Mercury 6, the first American manned spacecraft to go into orbit, lifts off from Cape Canaveral in Florida on February 20, 1962. The astronaut, John Glenn, spent nearly 5 hours in space and made 3 orbits of the Earth before splashing down safely in the ocean.

Below: The splashdown of the Skylab 2 mission. All American astronauts have returned to Earth in this way, their spacecraft descending gently into the ocean beneath parachutes. Recovery ships were waiting nearby.

Man in space

From the start of the space race, it was clear that Russia was aiming to put a man in space at the earliest opportunity. Many experiments were first made with animals. Several more dogs made spaceflights after Laika in Sputnik 2 and most, though not all, returned safely to Earth with no ill effects.

The development of a spacecraft with a spherical compartment large enough to take one cosmonaut (as a Russian astronaut is called) was soon completed and tested. On April 12, 1961, Yuri GAGARIN, a major of the Soviet air force, blasted off from TYURATAM spacedrome aboard the spacecraft VOSTOK 1. He landed back in Russia 108 minutes later, after one orbit of the Earth. A heat shield on the spacecraft bore the searing heat of re-entry, and the spacecraft was slowed by parachutes to land on the ground. Gagarin landed separately by parachute safe and sound.

The Americans called their manned space-flight programme the MERCURY PROJECT, and were cautious in their approach. A conical one-man space capsule was developed and tested, but the first manned flights did not go all the way into orbit. Two suborbital flights took place soon after Gagarin's triumph, and then an ape called Enos was sent into space to make two orbits in

---

Goddard Space Flight Center, at Greenbelt, Maryland, US, is a NASA installation named after the rocket pioneer Robert GODDARD. The Center is responsible for the tracking of and data acquisition from NASA's satellites and probes.

Gordon, Richard F. (1929–) is an American astronaut who flew on the Gemini 11 and Apollo 12 missions.

Grissom, Virgil I. (1926–67) was an American astronaut who made the second American sub-orbital flight in the MERCURY programme,

and flew aboard the first GEMINI mission. He was scheduled to be commander of the first manned Apollo flight, but he and his crew were killed in a fire on board the spacecraft during a practice countdown.

H **HEAO** is an abbreviation for High Energy Astronomical Observatory, a series of 3 US satellites intended to study X-rays and gamma rays from space. HEAO-1 was launched in 1977. The second and third craft are planned to follow in 1978 and 1979.

Kennedy Space Center

I **Intelsat** is the name of a series of satellites launched by the International Telecommunications Satellite Corporation, formed in 1964. The first Intelsat satellite was EARLY BIRD, launched in 1965. The latest series of satellites, called Intelsat IV-A, can carry approximately 6,000 telephone circuits or 20 colour TV channels. There are a total of 6 Intelsat IV-A satellites.

J **Jet Propulsion Laboratory** is a division of NASA at Pasadena, California. It is responsible for

NASA's deep-space tracking network, including the 64-metre antenna at Goldstone, California. Jet Propulsion Laboratory scientists are responsible for American planetary probes, such as the Mariner series.

Johnson Space Center is a division of NASA at Houston, Texas, responsible for directing manned space flights; the establishment was known as the Manned Spacecraft Center until 1973. JSC helps design spacecraft and train astronauts, and is the home of Mission Control.

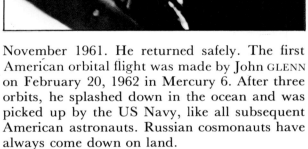

November 1961. He returned safely. The first American orbital flight was made by John GLENN on February 20, 1962 in Mercury 6. After three orbits, he splashed down in the ocean and was picked up by the US Navy, like all subsequent American astronauts. Russian cosmonauts have always come down on land.

Meanwhile a second Russian had gone into space. Gherman TITOV made 17 orbits in Vostok 2 in August 1961. The comparatively large number of orbits was necessary to return Titov to Russia. During the long flight he suffered from vertigo and nausea, although he returned to Earth unharmed. The Mercury and Vostok programmes continued until 1963, there being four orbital Mercury flights and six Vostok missions in all. The longest was Vostok 5, which lasted almost five days, and another Russian first was scored with Vostok 6 in 1963 when Valentina TERESHKOVA became the first woman to fly in space.

Multi-manned missions

The Vostok and Mercury programmes showed that people could fly safely in space and perform useful tasks, and that there was no immediate hazard from weightlessness, radiation or meteoroids. What direction was manned spaceflight now to take? The Russians declared no

Above left: The Gemini spacecraft was the first manned spacecraft that could manoeuvre in space. An historic event took place in orbit in December 1965, when 2 Gemini spacecraft made the first space rendezvous and came to within a metre of each other.

Above right: The first American spacewalk was made in June 1965 when Ed White, protected by his spacesuit, floated in space outside his Gemini spacecraft for 21 minutes. He was tethered to the spacecraft for safety, but manoeuvred himself by firing a hand-held gas gun.

objectives, as always, but the United States stated as early as 1961 that its intention was to land a man on the Moon by 1970.

The next phase was to send more than one man at a time into orbit, and here again the Russians got away first with the three-man VOSKHOD 1 mission in October 1964. This was followed by a two-man flight, Voskhod 2, in March 1965 and during this flight, Alexei LEONOV left the spacecraft in a spacesuit to make the first 'walk' in space. Only five days later, the first of America's two-man GEMINI flights was launched.

The Gemini spacecraft could manoeuvre itself through space, and in December 1965, two spacecraft were placed in orbit at the same time, Gemini 6 and Gemini 7. One spacecraft then altered its orbit to rendezvous with the other, a technique vital to the proposed Moon landings. Astronauts also made spacewalks to gain knowledge of how they would be able to work on the Moon. The Moon missions would also be lengthy, and Gemini 7 went on to make, for then, a record spaceflight of almost two weeks without harming the astronauts concerned. An anxious moment occurred on Gemini 8 when it began to tumble out of control, but the astronauts managed to right the spacecraft and returned to Earth early but safe.

K Kennedy Space Center is the NASA establishment at Cape Canaveral, Florida, responsible for preparing and launching manned and unmanned space missions. KSC houses the giant Vehicle Assembly Building in which manned rockets, such as the SATURN series and the SPACE SHUTTLE, are assembled.

Komarov, Vladimir M. (1927–67) was a Soviet cosmonaut and the first man to die during a space mission. Komarov was commander of the world's first 3-man flight, Voskhod 1, in October

1964. In April 1967 he test flew Soyuz 1, but was killed when the spacecraft went out of control and crashlanded.

Alexei Leonov in Soyuz

Korolev, Sergei Pavlovich (1906–66) was the designer of the first Soviet rockets and spacecraft. In 1929 he began to design aircraft but, inspired by TSIOLKOVSKY, turned to rocket-powered planes in 1932. He developed this work during World War II, and after the war headed the team that built the first Soviet rockets.

L Landsat is the name of 2 satellites launched in 1972 and 1975 to survey the Earth. The Landsats orbit from pole to pole, photographing the Earth with a

series of cameras to monitor crop growth and water and air pollution, and to prospect for possible new sources of minerals. Remote countries have also used Landsat photographs for making accurate new maps.

Leonov, Alexei A. (1934–) is a Soviet cosmonaut who was the first man to walk in space. During the Voskhod 2 mission in March 1965 he crawled through an airlock to spend 10 minutes in free space. Leonov commanded the Soviet crew during the Apollo-Soyuz linkup in July 1975.

Lovell, James A. (1928–) is an American astronaut who commanded the Apollo 13 Moon mission in April 1970. Its proposed lunar landing was cancelled after an on-board explosion that threatened the lives of the crew. Lovell first flew on the record-breaking Gemini 7 mission in December 1965, on Gemini 12 in November 1966, and on Apollo 8, the first manned craft to orbit the moon, in December 1968.

Luna spacecraft are a series of Soviet Moon probes. In September 1959

The Gemini programme ended with Gemini 12 in November 1966. It had been a hectic year and a half in space, but very successful. By contrast, the Russians made no more spaceflights after Voskhod 2. The Voskhod spacecraft was no more than a Vostok with extra seats, and a new craft had to be developed. The Americans were ready to press on with flights of the three-man APOLLO spacecraft, which were destined to fly to the Moon. At this promising point, tragedy struck both space nations. In January 1967, an Apollo spacecraft caught fire during a ground test, killing the three astronauts inside—Roger Chaffee, Virgil GRISSOM and Edward WHITE. And then two months later, Russia launched the first of its new SOYUZ spacecraft, only to have it crash on landing and kill its single cosmonaut, Vladimir KOMAROV.

Both the Soyuz and Apollo programmes were delayed by these tragedies. Both spacecraft eventually took to space in October 1968, Apollo

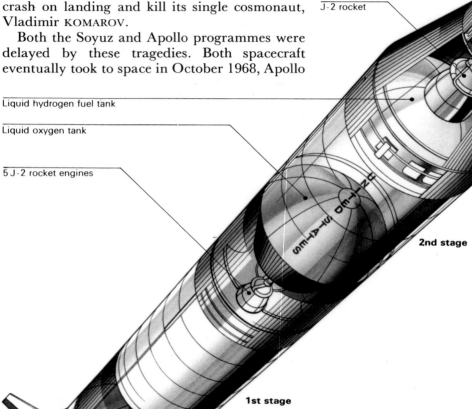

Emergency escape rocket
Apollo command module
Apollo service module
Lunar module
Helium bottles
J-2 rocket
Liquid hydrogen fuel tank
Liquid oxygen tank
5 J-2 rocket engines
2nd stage
1st stage
Kerosene fuel tank
5 F-1 rocket engines
3rd stage
Apollo
Saturn V (111 metres)
Jupiter C (22 metres)
Bus

Above: The Saturn V rocket that launched the Apollo spacecraft to the Moon. It is the biggest rocket yet to have been made.
Above right: The Saturn V compared in size to Jupiter C, the rocket that launched the first American satellite, Explorer 1, in 1958.

to prepare for the Moon and Soyuz to inaugurate an Earth orbital programme that continues today. The Russians had another disaster in 1971 with Soyuz 11, when an equipment failure killed the three cosmonauts during re-entry. The Americans had a very narrow escape with Apollo 13 in the previous year, but to date have lost no astronauts in space at all.

The way to the Moon
The path to the Moon had to be prepared by space probes. Close-up pictures of the Moon's surface and information about conditions on the Moon had to be obtained before sending astronauts to land there. Detailed Moon maps also had to be prepared to choose the best landing sites. The first close-up pictures were

Luna 2 became the first object to hit the Moon; in October 1959 Luna 3 sent back the first photographs from the Moon's far side; and in February 1966 Luna 9 sent back the first photographs from the Moon's surface. Several Luna probes, starting in September 1970 with Luna 16, have automatically returned small samples of Moon soil to Earth, and 2 Luna probes have delivered automatic Moon rovers (LUNOKHOD) to the surface.
Lunar Module was the craft in which 2 men touched down on the Moon during

the Apollo series of lunar landings. The Lunar Module's lower half contained a powerful descent engine to break the fall; the upper half

Docked module in space

separated from this at lift-off from the Moon. A separate ascent engine was used to boost the astronauts back up to the command module.
Lunar Orbiter was a series of 5 American Moon-circling probes that photographed the entire Moon, front and back, between August 1966 and late 1967. They showed details far finer than it was possible to see using telescopes on Earth, and helped scientists choose the safest sites for manned landings.
Lunokhod is the name of a Soviet automatic lunar roving vehicle, capable of

being driven from Earth by remote control. Lunokhod 1 was delivered to the Moon's surface in November 1970 by Luna 17, and Lunokhod 2 by Luna 21 in January 1973. The Lunokhods transmitted TV pictures of the lunar surface as they moved about.

Marshall Space Flight Center is a division of NASA at Huntsville, Alabama, where the SATURN family of rockets were developed by Wernher VON BRAUN. It is now responsible for development of engines for the SPACE SHUTTLE.

McDivitt, James A. (1929–) is an American astronaut who commanded the Apollo 9 mission in March 1969 which first

John Glenn

obtained by the American probe RANGER 7 in 1964 just before it crashed on the Moon. They showed a smooth surface marked with small craters. These pictures were not of great use, and the Russians began to take the lead in the development of automatic lunar exploration.

In February 1966, Luna 9 became the first space probe to soft-land on another world, and sent back pictures of the Moon's surface. They showed a hard rocky surface, and not the bowls of dust that some had suspected might swallow up anything landing on the Moon. Then only a month later, Luna 10 became the first probe to orbit the Moon, sending back pictures from which detailed maps could be made. The Americans soon followed. The SURVEYOR probes, beginning in May 1966, soft-landed to examine possible landing sites and a series of LUNAR ORBITER probes mapped the whole of the Moon from August 1966 onwards.

Meanwhile, special vehicles were being developed to take men to the Moon by a technique called lunar-orbit rendezvous. Astronauts could not fly directly to the Moon and back again, because a rocket of gigantic power would have been needed. The SATURN V rocket that America built to launch the Apollo spacecraft on its way was vast enough, far outclassing all other rockets to date and probably for some time to come. Its

Emergency oxygen tank

Emergency oxygen pipe

Back pack containing radio, oxygen and cooling equipment

Emergency oxygen switch

Emergency pressure and ventilation valve

Pressure gauge

Water-cooled underwear

Rubber pressure suit

Fibreglass covering

Gold-coated Sun visor

Backpack remote controls

Electrical cable

Oxygen pipe

Chronograph

Rock sample pouch

Utility pocket

Lunar overshoe

---

tested the Lunar Module in Earth orbit. McDivitt also commanded the Gemini 4 flight in June 1965 when Edward WHITE made the first American spacewalk.

Mercury project was the first US manned space programme. The conical Mercury capsule, 2.9 metres long and 1.9 metres wide, held 1 man. The first 2 Mercury flights, launched by Redstone rocket, were called suborbital as the spacemen, Alan SHEPARD and Virgil GRISSOM, did not go into orbit. Orbital Mercury flights, launched by ATLAS ROCKET, began

with John GLENN in February 1962 and ended with Gordon COOPER in May 1963.

N NASA, the National Aeronautics and Space Administration, is the US government agency in charge of civilian space programmes. NASA was founded in October 1958 with headquarters in Washington DC. Among NASA field stations are: AMES RESEARCH CENTER; GODDARD SPACE FLIGHT CENTER; JET PROPULSION LABORATORY; JOHNSON SPACE CENTER; KENNEDY SPACE CENTER; MARSHALL

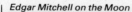
Edgar Mitchell on the Moon

SPACE FLIGHT CENTER; and Wallops Flight Center.

Nikolayev, Andrian G. (1929–) is a Soviet cosmonaut who in August 1962

flew in the Vostok 3 capsule, while Pavel POPOVICH was in orbit simultaneously in Vostok 4. Nikolayev flew again in Soyuz 9 in June 1970. In 1973 he married cosmonaut Valentina TERESHKOVA.

Nimbus was a series of 6 American weather satellites. They were used for developing new instruments to measure temperatures and humidity in the atmosphere and return cloud-cover photographs.

O OAO was a series of Orbiting Astronomical

Observatory satellites that observed the Universe at ultra-violet and X-ray wavelengths. OAO 1 was launched on April 8, 1966,

Typhoon seen from space

three stages took the spacecraft up into orbit and then the third stage fired it on its way out to the Moon.

The spacecraft itself consisted of three parts or modules and the three astronauts flew out to the Moon in the central COMMAND MODULE. Behind it was the service module, which contained the main engine and power supply. This engine fired the spacecraft into orbit around the Moon as it drew near. Then two astronauts transferred to the third section, the LUNAR MODULE. This module separated from the command module and descended on to the Moon, while the engine in the lower half of the module fired so that it could land on to the surface on its four legs.

After the astronauts had completed their various tasks on the Moon's surface, they returned to the command module in the upper half of the lunar module, using the lower half as a launching pad. They then made a rendezvous with the command module, still in lunar orbit, and rejoined the third astronaut aboard. The lunar module was then abandoned and the main engine fired to take the three crewmen back to Earth. Just before re-entry, the command module separated from the service module and then splashed down into the ocean beneath parachutes. Of the huge machine that left the Earth, only this tiny part returned, scorched by the heat of re-entry.

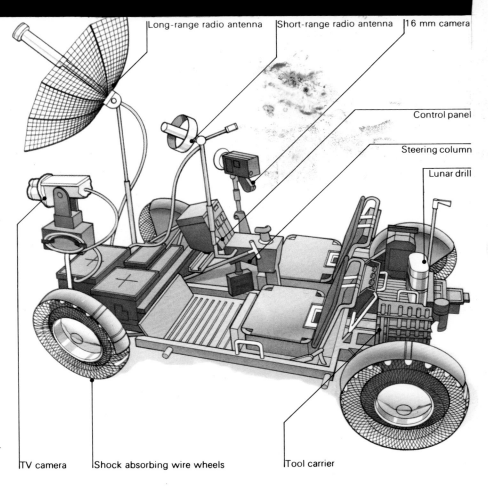

Long-range radio antenna Short-range radio antenna 16 mm camera

Control panel

Steering column

Lunar drill

TV camera Shock absorbing wire wheels Tool carrier

and OAO 2 on December 7, 1968. The most successful of the series was OAO 3, also known as COPERNICUS, launched on August 21, 1972. It carried an 81-cm reflector.
OSO was a series of 8 Orbiting Solar Observatory satellites, launched by the US between March 1962 and June 1975. The satellites studied the Sun at short wavelengths through a complete 11-year solar cycle.

P **Plesetsk** is a Soviet launch site near Archangel from which the majority of Soviet unmanned launches are made. Plesetsk, a former military missile site, made its first space launch in March 1966. Satellites from Plesetsk are used for weather monitoring, communications, and military reconnaissance.
Popovich, Pavel R. (1930–) is a Soviet cosmonaut who flew the Vostok 4 mission in August 1962, at the same time as Andrian NIKOLAYEV was in orbit in Vostok 3; this was the first simultaneous flight of 2 manned craft. In 1974 Popovich worked for 2 weeks in the Salyut 3 space

Saturn V liftoff

station with which he docked in Soyuz 14.

R **Ranger probes** were a series of American Moon probes. The first members of the series were meant to land an instrumented package on the Moon, but all failed. The probes were then made purely photographic. Rangers 7, 8 and 9 sent back detailed pictures of the Moon as they crashed into its surface in 1964 and 1965.

S **Salyut** is a Soviet space station capable of housing up to 3 men for several months. Salyut 1 was launched in April 1971. The Salyuts are regularly replaced by new space stations. Some of the Salyuts are used for astronomical and Earth observation, but others seem to be for military surveillance. Salyut is 12 metres long, weighs 18.5 tonnes, and has approximately 100 cubic metres of volume, making it considerably smaller than SKYLAB.
Saturn rockets were a family of large space launchers developed by Wernher VON BRAUN for manned space

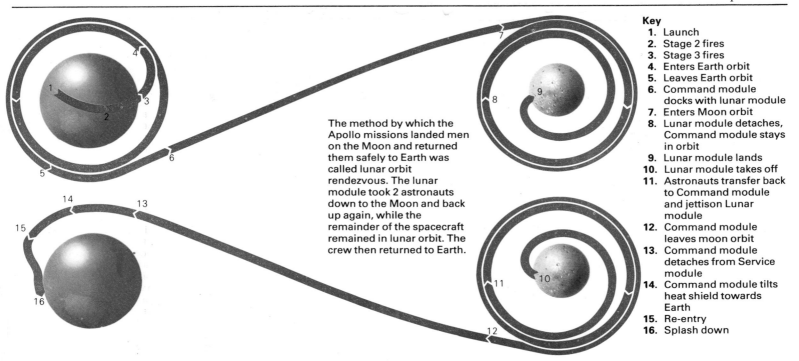

The method by which the Apollo missions landed men on the Moon and returned them safely to Earth was called lunar orbit rendezvous. The lunar module took 2 astronauts down to the Moon and back up again, while the remainder of the spacecraft remained in lunar orbit. The crew then returned to Earth.

Key
1. Launch
2. Stage 2 fires
3. Stage 3 fires
4. Enters Earth orbit
5. Leaves Earth orbit
6. Command module docks with lunar module
7. Enters Moon orbit
8. Lunar module detaches, Command module stays in orbit
9. Lunar module lands
10. Lunar module takes off
11. Astronauts transfer back to Command module and jettison Lunar module
12. Command module leaves moon orbit
13. Command module detaches from Service module
14. Command module tilts heat shield towards Earth
15. Re-entry
16. Splash down

Men on the Moon

After six unmanned tests, between October 1968 and May 1969, the Apollo spacecraft was tested on four manned missions, two in Earth orbit and two in orbit around the Moon. It functioned perfectly, clearing the way for the historic Moon landing mission of Apollo 11. The three astronauts chosen were Neil ARMSTRONG, Edwin ALDRIN and Michael COLLINS. Armstrong and Aldrin landed in the Sea of Tranquillity on July 20, 1969 and worked on the surface for 2.6 hours the following day. The work of all the Moon visitors was to collect samples of Moon dust and rock and to set up instruments. These measured such things as the Moon's seismic (earthquake) activity, heat and magnetism. The astronauts also set up mirrors to reflect laser beams from Earth, thus making acc▮▮▮te measurements of the distance to the Moo▮.

Five more landings ▮▮▮ place. Apollo 12 touched down on the ▮▮▮▮ of Storms, within walking distance of the space probe Surveyor 3, in November 1969. In February 1971, Apollo 14 visited a region near the Fra Mauro crater. Apollo 15, in August 1971, took in the most spectacular scenery of all the missions, landing in the foothills of the lunar Apennines and

Below: Working out the course of a journey to the Moon is complicated because the Earth and Moon are in constant motion. The spacecraft has to be fired to one side of the Moon, using the Earth's spin to gain speed (1—3), so that it will meet the Moon at a certain point (4) in its orbit around the Earth.

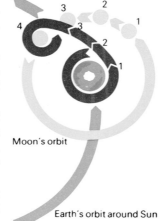

Moon's orbit

Earth's orbit around Sun

alongside the deep cleft of Hadley Rille. Apollo 16 landed in the lunar highlands in April 1972, and the programme came to an end with Apollo 17, which visited a valley in the Taurus Mountains in December 1972.

On the last three missions, the astronauts had the help of a Moon car to drive long distances from the lunar module. They covered up to 35 kilometres and spent as long as 22 hours in all on the surface. The scientific results of the landings have provided us with knowledge of the structure and origin of the Moon that could not have been obtained from Earth.

One mission, Apollo 13 in April 1970, never reached the Moon. An explosion wrecked the service module on the way, putting the main engine and power supply out of action. However, the crew were able to take to the lunar module, and use its life-support systems to survive and its engine to return them to Earth.

The Russians meanwhile contented themselves with automatic exploration of the Moon. Beginning with Luna 16 in September 1970, a series of probes have landed on the Moon, scooped up some soil and returned to Earth. Two Luna probes have also delivered automatic LUNOKHOD vehicles to the Moon.

---

uses. The smaller member of the family, the Saturn IB, was a 2-stage rocket used for launching Apollo capsules into Earth orbit. The larger rocket, the 3-stage Saturn V, was designed to send Apollo to the Moon. The Saturn IB was 64 metres high to the top of the Apollo command module, had a lift-off thrust of 740 tonnes, and could place 18 tonnes into orbit around the Earth. The Saturn V was 111 metres high, including the Apollo command module, had a lift-off thrust of 3,450 tonnes, and could place over 100 tonnes

into Earth orbit, or send over 40 tonnes to the Moon. These rockets have been made obsolete by the SPACE SHUTTLE.
Schirra, Walter M. (1923–) is an American astronaut who flew on the fifth Mercury mission in October 1962. He also flew on Gemini 6 in December 1965, which made the first rendezvous with another craft, Gemini 7. In October 1968 he commanded Apollo 7, the first manned flight of an Apollo. He thus became the only man to fly all 3 types of American manned spacecraft.

Scott, David R. (1932–) is an American astronaut who commanded the Apollo 15 mission which landed on the Moon in July 1971. He also

David Scott

flew on Gemini 8 in 1966, and on Apollo 9 in 1969.
Scout rocket is the smallest American satellite launcher, used to orbit small scientific satellites, including members of the ARIEL and EXPLORER series. Scout is a 4-stage rocket using solid propellants (a 5-stage version is also available).
Shepard, Alan B. (1923–) was the first American to fly in space. He made a suborbital ride in the Mercury capsule Freedom 7 on May 5, 1961. In February 1971 he commanded the Apollo 14 mission to the Moon.

Skylab was an American space station made from the converted upper stage of a Saturn V rocket. It was used to carry out research and to

Skylab

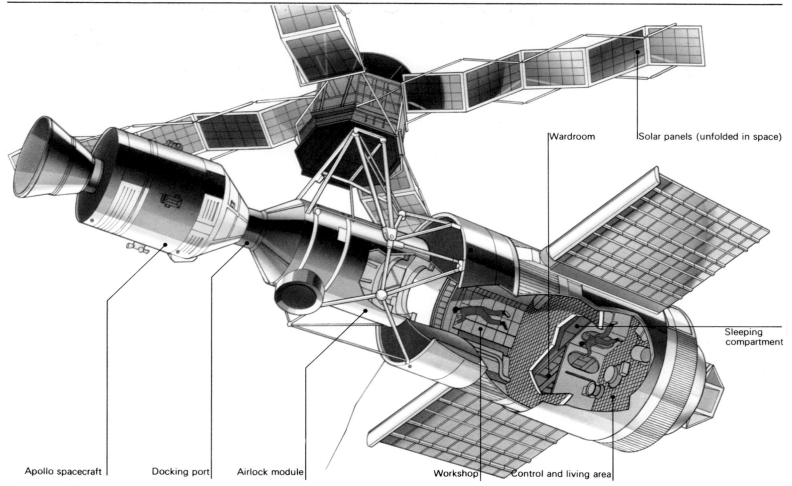

Wardroom | Solar panels (unfolded in space)

Sleeping compartment

Apollo spacecraft | Docking port | Airlock module | Workshop | Control and living area

Space stations

The Apollo project was an extraordinary success. To land men on the Moon only seven years after the first American ventured into space was a feat that will go down in history as one of the greatest human achievements. It was perhaps the most staggering demonstration of man's ingenuity that the world has ever seen. But although it marks our first steps to the stars, a lot remains to be done nearer home before we can venture farther into space.

A space station is a base where astronauts can work for long periods. The astronauts use their spacecraft only as ferries to take them to and from the space station, in which they can live and work in comfort. The space station may be launched into orbit by rocket, or it may be assembled by astronauts from pieces sent up by rocket. Missions to the Moon and planets could leave from space stations and return to them

Above: The Skylab space station was the largest spacecraft that man has ever constructed. It orbited the Earth and was visited by teams of 3 astronauts on missions that lasted as long as 84 days. The crew has separate living and working quarters. Power came from panels of solar cells, some of which the astronauts had to repair. The Skylab missions made important observations of the Sun and the Earth, and carried out experiments in the weightless conditions aboard. They also showed that people can survive long spaceflights.

instead of the ground. The interplanetary spacecraft could then be assembled at the space station so that they would not have to be built to withstand the strains of lift-off from the ground and re-entry into the atmosphere.

It has also been suggested that our energy problems could be solved by orbiting huge constructions to capture the Sun's energy. They would convert this energy into beams of microwaves that would be picked up by ground stations and turned into electricity.

It is unlikely that human beings will be able to take much part in exploring the planets; unmanned space probes are physically much more adaptable than the human frame. However, space stations will demand the agility of the human mind.

Firstly, a space station can function as a workshop and laboratory. In the weightless conditions aboard, experiments can be made that

observe the Earth, Sun and other stellar objects. Skylab was launched on May 14, 1973, and was later manned by 3 crews, each of 3 men, who stayed for up to 84 days. Skylab was 36 metres long overall, had a total internal volume of 350 cubic metres, and weighed 75 tonnes, far larger than the Soviet SALYUT. The last crew left Skylab in February 1974, and there are no plans to reoccupy it.
Small Astronomy Satellites are a series of satellites in the Explorer series to study X-rays and gamma

rays from the Universe. SAS-1 (also known as Explorer 42 and Uhuru) was launched on December 12, 1970, to survey the X-ray sky. SAS-2, launched on November 16, 1972, studied gamma rays. Last of the series was SAS-3, launched in May 7, 1975 for further detailed studies of X-ray sources. Work of the SAS satellites has been taken over by the HEAO series.
Soyuz is a Soviet manned spacecraft, first launched in April 1967. Soyuz consists of a bell-shaped crew compartment 2.2 metres long and

wide, which contains 2 cosmonauts in spacesuits, with a spherical work compartment in front of it. The work compartment and the service module, which contains a manoeuvring engine and solar panels, are jettisoned before the crew compartment returns to Earth. Soyuz

Russian spacecraft Soyuz

can be used for independent missions or to ferry cosmonauts to SALYUT space stations. A total of 26 Soyuz missions had been flown by the end of 1977.
Space Shuttle is a reusable space plane for ferrying satellites and men into orbit. The winged Shuttle is boosted into orbit by 2 solid-fuel rockets which drop away and are recovered. It also has its own engines which are fuelled from a giant tank strapped to the Shuttle's belly. The length of the Shuttle orbiter is 37.2 metres; its cargo bay is 18.3

may lead to new manufacturing processes. Perfect lenses could be made in conditions where gravity would not act to distort them. So could perfect ball bearings – which are impossible to achieve on Earth for the same reason – and perfect crystals, which could revolutionize electronics. Because lighter materials do not float on heavier materials in weightlessness, better alloys could be made in which the components do not tend to separate out. Experiments can also be made in the vacuum of space, a vacuum far rarer than any that can be obtained on Earth.

Space stations can also act as observatories – both of the skies above and the Earth below. Satellites function in this way at present, but human observers can make better use of space by deciding when and where to point their instruments. Furthermore, as well as constructing and assembling space vehicles and hardware, astronauts aboard a space station could also repair them.

The earliest steps towards a permanent base in space were taken by the Russians. The first semi-permanent space station, called SALYUT 1, was launched into orbit in April 1971. Salyut was much larger than any other previous spacecraft and contained a host of equipment to carry out the kinds of experiments and observations described above. It received its first visitors only four days after launch, when the spacecraft Soyuz 10 docked with Salyut. However, the crew did not enter the station. This great space achievement was left to the three-man crew of Soyuz 11, who in June 1971 stayed aboard in space 24 days (a record at that time). Then tragedy struck. The three cosmonauts left Salyut without trouble, but were found to be dead when their Soyuz spacecraft was opened after it had landed.

Doctors feared that weightlessness might be to blame. Weakened by a long period without gravity, the cosmonauts might not have been able to withstand the strains of re-entry. If this were the case, long spaceflights would be out of the question. However, the cause was found to be an equipment failure. Other technical problems have since dogged the Salyut programme, though none so serious, but it has gone ahead steadily. By the end of 1977, six Salyut stations had been placed in orbit. The longest stay had been of 63 days aboard Salyut 4 in 1975.

Above: All astronauts undergo a long training programme. To simulate weightlessness, the trainees are taken on aircraft that fly in special manoeuvres that make everything aboard lose gravity for a short time.

Below: The astronauts on the Skylab missions spent many hours observing the Earth below and taking photographs of interesting features. This photograph is of Long Island in the United States.

metres long by 4.6 metres wide, and can carry up to 29.5 tonnes into orbit. Because the Shuttle orbiter flies back to land like an aircraft, it is anticipated that it can be reused 100 times or more. It could thus cut the cost of launches by up to 90%. The first Shuttle launches are planned for 1979, with as many as one flight a week by the 1980s.
Space telescope is a 2.4-metre reflecting telescope to be launched by the Space Shuttle in the 1980s. Such a telescope would revolutionize astronomy by

enabling us to see objects 100 times fainter than those visible from the ground and in 10 times as much detail.
Spacelab is a space station built by the European Space Agency to fly in the cargo bay of the Space Shuttle. Inside Spacelab astronauts will be able to perform experiments in weightlessness and observe the Earth and sky for up to a month at a time. The first Spacelab mission, in 1980, will include a European astronaut.
Sputnik was the name given to the first series of Soviet Earth satellites. Sput-

nik 1, launched on October 4, 1957, was the first artificial Earth satellite. Sputnik 2, launched on November 3, 1957, was the second and

Artist's impression of a space colony

carried the dog Laika. Sputnik 3 launched on May 15, 1958, was a scientific satellite that confirmed the existence of the VAN ALLEN BELTS

(*see page 47*), discovered by EXPLORER 1. Later Sputniks (the series ended with Sputnik 10 in March 1961) were test flights of the VOSTOK capsule. The Sputniks were superseded by the COSMOS series.
Stafford, Thomas P. (1930-) is an American astronaut who commanded the Apollo spacecraft in the Apollo-Soyuz mission in July 1975. Stafford also flew in Gemini 6 and Gemini 9, and in 1969 commanded Apollo 10, the dress rehearsal for the first manned lunar landing.
Surveyor probes were a

The Americans have not really tried to compete. They have orbited only one space station, called SKYLAB, and it was made by converting the third stage of a Saturn rocket. However, the resulting station was huge – nearly four times the volume of Salyut. Three crews, each of three astronauts, visited Skylab in Apollo spacecraft between May 1973 and February 1974. Although the station was damaged at launch, the astronauts were able to repair it and the three missions were very successful. Skylab's main achievement was in solar astronomy. It was equipped to make observations of most of the radiations produced by the Sun, many of which do not penetrate the atmosphere. The missions gave astronomers more information about the Sun than had been achieved in centuries of Earthbound observation.

Prolonged weightlessness was generally expected to be a problem aboard space stations. The Skylab astronauts took fish and spiders with them to see how they would cope away from gravity, but they adapted very quickly. The astronauts had to exercise often to keep their bodies in good shape. Strangely enough, the third crew, which stayed a record 84 days in space, adapted to weightlessness better than the previous two crews and returned in excellent

Right: The Space Shuttle is launched by 2 booster rockets, which are later recovered. It goes on up into orbit using fuel from a fuel tank that is jettisoned and not recovered. After its mission is completed, the shuttle re-enters the atmosphere and glides down to land on a runway like a conventional aircraft. It can then be used again for further missions.

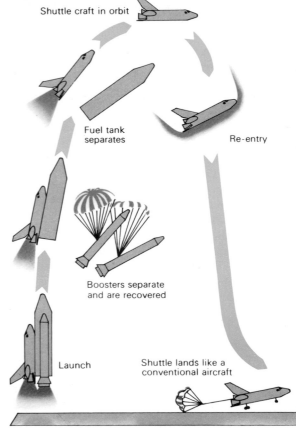

Shuttle craft in orbit

Fuel tank separates

Re-entry

Boosters separate and are recovered

Launch

Shuttle lands like a conventional aircraft

Below: The first tests of the Space Shuttle took place in 1977. A shuttle was carried aloft fixed to the back of a jumbo jet, and then released to glide back to the ground and land on its own.

series of American lunar soft-landings that paved the way for eventual manned landings. (Soft-landings are controlled by some form of deceleration, such as retro-rockets.) Last of the series was Surveyor 7 which landed on January 10, 1968. Although 2 of the series were failures, the 5 successful Surveyors returned photographs from the lunar surface and performed remote chemical analyses of the lunar soil.
Synchronous orbit is an alternative term for a GEOSTATIONARY ORBIT.

Telstar was the name of two American communications satellites, launched on July 10, 1962 and May 7, 1963 respectively. Telstar 1

Space freighter from the film Star Wars

carried the first live Transatlantic TV, and both Telstars linked Europe and the US by TV and telephone.
Tereshkova, Valentina V. (1937-), Soviet cosmonaut, was the world's first (and so far only) woman to fly in space. She orbited the Earth 48 times in Vostok 6 in June 1963. She later married the cosmonaut Andrian NIKOLAYEV.
Tiros was a series of US weather satellites. Tiros 1, launched on April 1, 1960, was the world's first weather satellite. The series ended with Tiros 10 in July 1965, and has been superseded by a more advanced system. The series returned over half a million photographs of the Earth's cloud cover.

Titan rockets are a family of American space launchers used for manned and unmanned missions. The 2-stage Titan II, based on an intercontinental military missile, was used to launch GEMINI spacecraft. More advanced versions, known as Titan III, are used to launch satellites, and planetary probes such as VIKING and VOYAGER (*see page 48*).
Titov, Gherman S. (1935-) was the second Soviet cosmonaut to fly in space. In August 1961 he made 17 orbits of the Earth, lasting over a day.

condition. Weightlessness is not therefore likely to be a medical problem. However, it often caused the Skylab astronauts to lose their orientation, and things that were 'dropped' floated away in mid-air and were often lost. The astronauts sometimes found themselves drifting helplessly in mid-air and without help, it took as long as 20 minutes to reach the other side of the cabin! Future space stations may be rotated as they orbit, so that centrifugal force will hold the astronauts to the outer walls and create an artificial gravity inside.

There and back again

In all the manned spaceflights to date, the situation has never arisen where something has gone wrong and the crew could only be saved by sending up a rescue vehicle. This was just as well, for it was very unlikely that one could have been made ready in time. Steps were taken to remedy this state of affairs with the joint Apollo-Soyuz mission of July 1975. The American and Russian spacecraft met in Earth orbit and docked with one another. In future, one nation may be able to rescue another in space.

Apart from the joint flight, the United States has made no more manned missions since the Skylab programme ended. The Russians continued the Salyut programme, and by the end of 1977 had made a total of 34 manned spaceflights against the American total of 29. The 1980s will probably see further development of the SPACE SHUTTLE programme.

Until now, spaceflight has been very costly because it has been extremely wasteful. Each manned mission has used up a new rocket and spacecraft – as if one had to buy a new car every time one wanted to make a journey. If manned spaceflight is to have any future, reusable spacecraft will have to be built. The space shuttle is the first of a new generation of spacecraft. It looks like an aircraft, but it is launched vertically into space like previous spacecraft. At launch, it has two large booster rockets and an external fuel tank bigger than the shuttle itself fixed to its belly. The shuttle's rocket engine and the booster rockets fire to launch the spacecraft. The boosters take it up to 50 kilometres and then drop away, to land beneath parachutes and be recovered for future use.

The shuttle goes on into orbit, using fuel from

Above: The space station from the film *2001: A Space Odyssey*. The station rotates so that centrifugal force produces artificial gravity at the outer walls. A future space station would probably be of similar design.

the external tank. The tank is then jettisoned, and is not recovered. The shuttle then manoeuvres itself into an orbit up to 1,000 kilometres high and remains there for as long as a month. Its mission completed, it drops back to Earth and glides down to land on a runway, just like an aircraft. Each shuttle should be capable of about 100 flights, and manned spaceflight will become routine with a flight every week.

Tests of the shuttle began in 1977 and the first manned flight should take place in 1979. The shuttle's duties will be to place satellites in orbit, and to repair and retrieve them. It will also set up SPACELAB, a space laboratory being built by the EUROPEAN SPACE AGENCY, which will be manned by European as well as American astronauts. A permanent space station will also become a possibility as the shuttle could construct and service it. And with the shuttles standing by, no astronaut should fear being stranded in space.

The future in space

The space shuttle may prove that space has a commercial value: that special products, such as microelectronics, can be manufactured in orbiting workshops on an economic basis. The

---

Tsiolkovsky, Konstantin E. (1857–1935) was a Soviet pioneer of spaceflight, who in the late 19th century worked out the theory of rocket propulsion and predicted the launching of artificial Earth satellites. Tsiolkovsky showed how multistage rockets would be needed to break away from Earth, and even suggested fuels, such as liquid hydrogen and liquid oxygen, which are widely used today. Tsiolkovsky never built rockets himself, but he inspired the Soviet interest in spaceflight.

Tyuratam is the location north-east of the Aral Sea of the main Soviet launch site. It is also known as Baikonur which is actually a town

Valentina Tereshkova

nearly 300 km away. Tyuratam is the Soviet equivalent of Cape Canaveral; from here are launched all manned missions and space probes. The first Earth satellites were also launched from here, though more satellites are now launched from PLESETSK than from Tyuratam.

U **Uhuru** is the popular name for the first Small Astronomy Satellite, SAS-1, launched in December 1970 to study the X-ray sky. Uhuru is the Swahili word for freedom.

V **Vanguard** was a series of early American Earth satellites intended to put the first objects into space but eventually beaten by SPUTNIK and EXPLORER. The first Vanguard launch attempts in late 1957 and early 1958 were failures. Vanguard 1 was launched on March 17, 1958, and was followed by two others in 1959.
von Braun, Wernher (1912–77) was an American rocket engineer, born in Germany, who developed the family of SATURN ROCKETS for manned space flight. In Germany during World War II,

von Braun designed the V2 rocket. After the war, von Braun moved to the US, where a derivative of the V2, the Juno 1, was used to

Skylab's spider

harnessing of the Sun's energy by orbiting power stations could be another way in which space exploration will be of value to us all. In these ways, we may gain a practical reward from space. However, a prime objective of space exploration will still be the furthering of knowledge. We occupy but a tiny corner of the Universe, and all kinds of secrets must lie in wait for us beyond once the problems of interstellar travel are solved.

It is impossible to say what the heavens hold in store for us, but we can make a reasonable guess at our space activities in the near future. By the year 2000, space probes should have made a fairly thorough examination of the whole Solar System, with the exception of the most distant planet, Pluto. We should have a good idea of the nature of the other planets and some of their moons, and of such bodies as asteroids and comets. Manned flights will probably also have continued to the Moon, and a base may even

Below: A space battle from the film *Star Wars*. Unlike *2001*, which forecasts how space travel may develop in the near future, *Star Wars* is almost total fantasy. However, several features of the film may be prophetic. It has been reported that 'killer' satellites exist, capable of crippling spacecraft with laser weapons in much the same way as these space fighters.

have been set up there. On the Moon, optical astronomers would not face the hindrance of an atmosphere, and radio astronomers would benefit greatly from an observatory built on the Moon's far side, away from radio interference from Earth. Such a venture should be an international one following the first cooperative space missions of the 1970s and 1980s. In addition, there could be advanced plans for a manned mission to Mars.

Further speculation could take us into the realms of fantasy. Yet there is one certain event lying many millions of years ahead for the human race. As the Sun dies, it will swallow up the Earth. If we are to survive indefinitely, we must eventually find some way of travelling to the stars. Our first steps into space may mark the start of this long journey. Just as our scientific endeavours could bring about our imminent destruction, so they could also mark the beginning of our ultimate survival.

WORLD OF KNOWLEDGE

The Earth

Peter Harben

Introduction

In the 1960s, an old idea, the theory of continental drift, finally gained respectability as more and more research indicated that the face of our planet Earth was constantly changing. The theory that the continents have moved, and are still moving, has had enormous repercussions. It has helped us to understand many of the Earth's phenomena, including earthquakes, mountain building and volcanic activity. It has even led to a re-appraisal of the study of the evolution of life on Earth. **The Earth** includes a broad survey of the Earth sciences today, taking full account of the revolutionary theory of continental drift and other modern ideas on mineralogy, climatology, oceanography and geomorphology. It also examines the Earth as a habitat for mankind and warns us that global disaster could occur should we disturb our environment and squander the natural resources on which we all depend.

Early superstitions about the Earth have been gradually superseded by more sophisticated ideas. Modern scientists are beginning to understand the evolution of our planet and the dynamic forces which are still changing it.

Understanding the Earth

Early man explained the creation of the Earth and all its natural occurrences through stories – myths and legends – of giant beasts or powerful gods. According to the ancient Greeks, volcanoes erupted due to the activities of Hephaestus, the blacksmith god of the underworld, lighting his forge. The ancient Greeks also thought that Pluto, the king of the underworld, and his three-headed dog, Cerberus, guarded the mineral wealth within the Earth.

Many other peoples before and after the ancient Greeks had their own ideas about how the Earth functions. The Romans adopted many of the Greek gods, and renamed them. Hephaestus became the Roman god Vulcan, from whom the word volcano originates. The Indians had Devi, a god with a dual role of both mother Earth (protector) and the dark forces of the Earth (destroyer). In Scandinavian mythology, dwarfs were the owners of the mineral wealth in the ground. According to a Celtic legend, Lough Neagh in Ireland was formed by a giant pulling out a piece of the Earth. He flung it into the sea and created the Isle of Man.

Catastrophism

The great Flood occurs in many legends, from as widely-separated peoples as the Assyrians and Babylonians of south-western Asia to the Incas of South America. The story of Noah's Flood is told in the Bible.

Belief in the Flood developed in Europe in the Middle Ages into a theory that the Earth's features, including mountains and valleys, had been formed during sudden, violent catastrophes. This theory was called catastrophism. Outside Europe, some scholars had other ideas. For example, an Arab scholar, Avicenna (980-1037), argued that the Earth's surface was changed slowly by the action of running water. But such ideas did not find favour in Europe.

Above: This map of the world, prepared in 1154 by the Arab scholar al-Idrisi, shows a flat earth, enclosed by water. It was the most advanced map of the world in its day. (The south is at the top.)

Left: According to Hindu legends, the Earth rested on a golden plate, supported on the backs of elephants. In turn, the elephants stood on a giant turtle, representing the god Vishnu. Earthquakes were thought to occur when the elephants moved.

Reference

A **Agricola,** Georgius (1494-1555) was the Latin name taken by the German, Georg Bauer, who is often called 'the father of mineralogy'. He was the first to describe minerals, how they occurred and how they were mined in his day.

Amber is a hard, yellowish substance formed from plant resin. It makes beautiful jewellery. Insects may be trapped inside amber as FOSSILS.

Georgius Agricola

B **Basalt** is the most common IGNEOUS ROCK which solidifies on the Earth's surface. There are many kinds of basalt. They are fine-grained, dark grey or black in colour, and contain comparatively little silica. Basalt makes up most of the ocean floor.

The Giant's Causeway

C **Cenozoic era,** the most recent in Earth history, covers the last 63-64 million years and includes the TERTIARY and QUATERNARY periods. Cenozoic means 'new life'.

Cambrian period (570-530 million years ago) is the first period of the PALAEOZOIC ERA and Cambrian rocks are the earliest in which FOSSILS are abundant. *Cambria*, in Latin, means Wales, the area in which rocks of this period were first studied.

Carboniferous period lasted from 345 to 280 million years ago. In North America, it is divided into the Mississippian and Pennsylvanian periods. Much coal (largely composed of carbon) was formed during the Carboniferous period.

Cleavage, a property of some minerals and rocks, is a line of weakness along which substances can be split into thin slices. Slate and mica have good cleavage.

Coal, formed mostly during the CARBONIFEROUS PERIOD, is a SEDIMENTARY ROCK composed largely of carbon. It was formed from decomposed plants which were com-

The dawn of Earth Sciences

In the 1400s, many people believed that the rocks covering the Earth had been formed during the Flood. But LEONARDO DA VINCI, an Italian artist and scientist, rejected this simple explanation. He suggested that FOSSILS in rocks were the remains of ancient animals and plants. He noted that fossils often occurred in several rock layers which were separated by barren layers. This led him to think that the rocks had been formed in several calm episodes, rather than in one violent deluge.

Most people also believed that the Earth was of recent origin. In 1654, the Irish Archbishop James Ussher (1581–1656) worked out, from Biblical history, that the Earth had been created at 9 AM on October 26, 4004 BC!

In the late 18th century, an important debate began concerning the origin of rocks. A German professor, Abraham WERNER, suggested that rocks were formed from chemicals in sea-water.

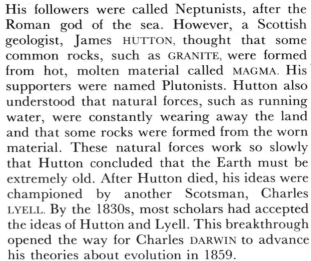

Above: The diagram shows how rocks such as granite are formed from molten material called magma. James Hutton and his followers, the Plutonists, pioneered this view in the late 1700s. Another group called the Neptunists argued that granite was formed from chemicals that had been dissolved in water.

Below: This space photograph shows a modern view of our planet, with Africa at the top.

His followers were called Neptunists, after the Roman god of the sea. However, a Scottish geologist, James HUTTON, thought that some common rocks, such as GRANITE, were formed from hot, molten material called MAGMA. His supporters were named Plutonists. Hutton also understood that natural forces, such as running water, were constantly wearing away the land and that some rocks were formed from the worn material. These natural forces work so slowly that Hutton concluded that the Earth must be extremely old. After Hutton died, his ideas were championed by another Scotsman, Charles LYELL. By the 1830s, most scholars had accepted the ideas of Hutton and Lyell. This breakthrough opened the way for Charles DARWIN to advance his theories about evolution in 1859.

Geology comes of age

By 1800, geology had become a respectable science. The first geological society was set up in London in 1807. A British surveyor and engineer, William SMITH, now made a vital contribution. He realized that, in rocks which have not been disturbed, the more recent rocks always overlay the older rocks. Smith was also interested in the fossils in the rocks. Some fossils appeared in several layers but some, called index or zone fossils, occurred in only one layer. Hence, if rocks, many kilometres apart, contain the same index fossils, then they must have been formed at the same time. Applying these principles, Smith produced a geological map of southern England, the first of its kind, in 1815.

Geological mapping continued and geologists identified and named rocks according to their relative ages. Many British names are used because most of the early work was carried out in Great Britain. Geologists eventually divided the history of the Earth into a few long eras. Eras were sub-divided into periods and some of the periods into epochs.

Various estimates of the Earth's age were made in the 19th century. For example, the British physicist Lord Kelvin (1824-1907) argued that the Earth must be between 20 and 30 million years old. But it was not until the discovery of radioactivity in the early 1900s that scientists were able to find the absolute ages of rocks and to estimate, accurately, the Earth's age (see page 74).

---

pressed in layers. Lignite (brown coal) contains up to 50% water. Bituminous coal contains much less water and anthracite, the hardest coal, contains very little water.
Core, Earth's. The Earth's core has a diameter of about 6,940 km. The inner core is solid, but the outer core is liquid.
Cretaceous period lasted from 136 to 63-64 million years ago. *Creta* is the Latin word for chalk and, in Europe and parts of North America, great thicknesses of chalk were formed in this

period. Dinosaurs became extinct at the end of the period.
Crust, Earth's. The Earth's crust is the hard outer skin of

Quartz crystal

the Earth. The thickness varies between 60-70 km under continental mountain ranges, to 6 km on the ocean floor. The crust ends at the Mohorovicic DISCONTINUITY.
Crystals are the natural shapes of minerals if they are allowed to form without interference. They are used to identify minerals.

D **Darwin,** Charles (1809-1892) was the British scientist who developed the theory of evolution by natural selection, part of which was based on FOSSIL evidence.

Dating rocks, see page 74.
Devonian period lasted from 410 to 345 million years ago. It is often called the 'age of fishes'. At the end of the period, amphibians first evolved from air-breathing fishes. The name Devonian comes from the English county of Devon.
Discontinuity is a boundary between different zones of the Earth's interior. The most important are the Mohorovicic (or Moho) between the crust and the MANTLE, and the Gutenburg, dividing the CORE from the mantle. Discontinuities were

detected by measuring the differences in density of each zone by the velocity of shock waves passing through them.

E **Elements** are the simplest forms in which substances can exist naturally. They combine to form compounds. Over 100 elements have been identified. The most common in the Earth's crust are oxygen and silicon, which together make up 74% of its weight.

F **Fossils** are evidence left by once-living organisms

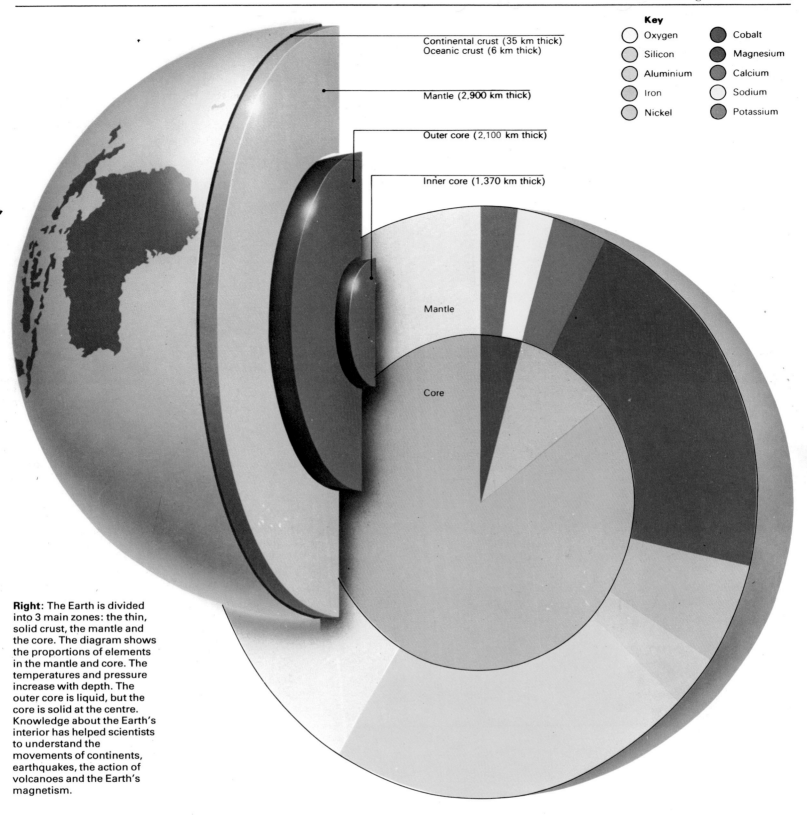

Continental crust (35 km thick)
Oceanic crust (6 km thick)

Mantle (2,900 km thick)

Outer core (2,100 km thick)

Inner core (1,370 km thick)

Key

○ Oxygen   ● Cobalt
○ Silicon   ● Magnesium
○ Aluminium   ● Calcium
○ Iron   ○ Sodium
○ Nickel   ○ Potassium

Mantle

Core

Right: The Earth is divided into 3 main zones: the thin, solid crust, the mantle and the core. The diagram shows the proportions of elements in the mantle and core. The temperatures and pressure increase with depth. The outer core is liquid, but the core is solid at the centre. Knowledge about the Earth's interior has helped scientists to understand the movements of continents, earthquakes, the action of volcanoes and the Earth's magnetism.

Ichthyosaur

in rocks. They have been preserved, often for millions of years, by natural processes. This may be because parts of organisms, such as bone or teeth, are hard, or because creatures were frozen in ice or trapped in AMBER. More commonly, parts of organisms were replaced by minerals, forming stone replicas of the originals. Other fossils include imprints, casts or carbon smears, such as those made by leaves. Fossils are useful in identifying rocks.

G **Geography** is the study of the Earth's surface, including physical features, vegetation, climate, soils, the oceans and the distribution and ways of life of people.

Geology deals with the origin, structure and history of the Earth and the study of the rocks in the Earth's crust.

Granite is a common IGNE-OUS ROCK, usually formed from large, underground masses of molten MAGMA. It occurs on the surface when cover rocks are worn away.

H **Hardness** of rocks is measured by such scales as Mohs scale which is numbered from talc (1), the softest, to diamond (10), the hardest.

Heat gradient is the mea-

Ptolemy's map of the world

Inside the Earth

To understand the ways in which natural phenomena occur, scientists had to find out about the nature of the Earth's interior. The Earth contains a thin CRUST enclosing the MANTLE, which surrounds the CORE. Scientists think that the core consists of two parts. The inner core is solid, but the outer core is liquid.

How was this picture of the Earth's interior arrived at? No one has ever drilled through the thin crust to find out what lies below. One attempt, the MOHOLE project, was abandoned by the US government in 1966 because of the cost.

Our knowledge of the Earth's interior has come from several sources. Some interesting evidence arrives from space in the form of meteorites. Some meteorites are composed of heavy, metallic iron-nickel material. Others are lighter, non-metallic and stony. Scientists have suggested that the heavier material may come from the dense core of a planet, while the lighter material may come from a planet's crust.

Evidence from earthquakes

Earthquakes create shock waves, just as stones create ripples in a still pond. These waves pass through the Earth's interior and can be recorded. There are two main kinds of waves. Primary (P) waves travel equally well through both solid and liquid rocks. Secondary (S) waves travel only through solids. At the same time, the speed of the waves varies according to the density of the rocks through which they are passing.

By measuring and tracking shock waves, scientists have built up an 'X-ray' picture of the Earth's interior. Measurements have shown that the upper mantle is composed of rocks with a density of about 3·4 grams per cubic centimetre. This is much denser than the continental crust, whose density is 2·7 and the oceanic crust (density 3·0). Geologists believe that, at the right temperatures and pressures, rocks such as peridotite and olivine could produce the higher density of the upper mantle. Minerals which are common in the lighter crust, such as quartz and feldspar, must be almost totally absent.

Other evidence

Scientists believe that some mantle rocks have been squeezed up into the crust. For example, peridotite, probably from the mantle, has been

Above: Mount Olympus is the highest mountain in Greece. The ancient Greeks regarded it as the home of the gods.
Below: The diagram shows the paths of shock waves, caused by earthquakes, through the Earth's interior. By tracking the paths and speeds of the waves, which vary according to the pressure, scientists have built up a picture of the Earth's interior.

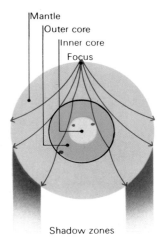

Mantle
Outer core
Inner core
Focus

Shadow zones

found in four locations, notably on Mount Olympus in Cyprus. The other locations are to be found in the Alps, Saudi Arabia and Papua New Guinea. The lava that erupts from Hawaiian volcanoes is believed to originate between 50 and 60 kilometres below the islands. Study of the lava confirms the theories about the mantle's composition.

The density of the mantle increases downwards from 3·4 to 4·5 grams per cubic centimetre. The overall density of the Earth, however, is known to be 5·5. The Earth's core must, therefore, be incredibly dense, between 10 and 13. This density is possible if the core is composed of nickel-iron, as suggested by the evidence of meteorites.

This theory is also supported by magnetic evidence. The Earth is like a giant magnet, with two magnetic poles. The electric currents needed to generate the Earth's MAGNETIC FIELD require a good electrical conductor, such as iron-nickel. Scientists also believe that the driving force of the magnetic field is supplied by the relatively faster movements of electrons in the liquid outer core. This electron movement creates a natural dynamo. These movements are probably caused by the rotation of the Earth.

---

sure of the increase in temperature as one descends into the Earth. The average rate is 1°C per 30 to 35 metres.
Hutton, James (1726-97), a Scottish farmer, became one of the greatest early geologists. He realized that rocks were continually being formed and destroyed and that the Earth was of great age. He also recognized that many rocks were igneous. His followers, called Plutonists after Pluto, god of the underworld, debated with, and triumphed over, the Neptunists, followers of Abraham WERNER.

I **Ice ages** are periods of Earth history when ice covered areas that are now ice-free. In the recent Pleistocene Ice Age, ice covered

Engraving from Hutton's 'Theory of the Earth'

much of the Northern Hemisphere. In the late CARBONIFEROUS and early PERMIAN periods, an ice age occurred in the Southern Hemisphere.

Igneous rocks are formed from molten MAGMA. The word igneous comes from the Latin *ignis*, meaning 'fire'. Some igneous rocks are intrusive (forming below the Earth's surface) and some are extrusive (forming on the surface). They are classified according to their quartz content, grain size, texture and so on.

J **Jurassic period** was a time when dinosaurs roamed the Earth. The period extended between 190-195 and 136 million years ago.

K **Kimberlite** is an IGNEOUS ROCK formed deep within the Earth. It occurs in long, narrow pipes and often contains diamonds.

L **Lava** is molten MAGMA which erupts from volcanoes or from cracks in the ground.
Leonardo da Vinci (1452-1519) is best known as an artist, but this Italian was also a scientist and inventor. He recognized that FOSSILS were the remains of once-living organisms, but rejected the belief that they were parts of creatures des-

The origin of the Earth's zones

About 5,000 million years ago, the Sun and the planets, including the Earth, formed from a revolving mass of hot dust and gas. For about 1,000 million years, lighter materials, such as carbon, oxygen, silicon and so on, gradually moved to the surface of the Earth. At the same time, heavier materials sank towards the centre.

Gases were released from the rocks by tremendous volcanic activity. These gases formed the first atmosphere, probably consisting largely of hydrogen and helium. But most of this primitive atmosphere probably escaped into space. Our modern, oxygen-rich atmosphere only began to form about 2,000 million years ago, after oxygen-producing plants had become established.

Elements and minerals

Everything in the Universe is made up of elements (substances that cannot be broken down into anything simpler). Scientists have identified over 100 elements, but more than 98

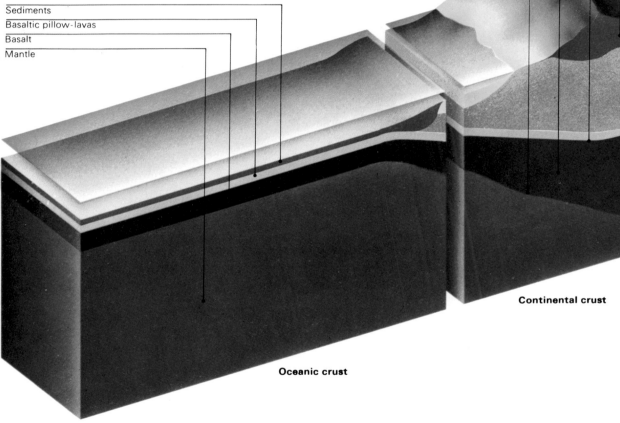

Left: The chief elements in the Earth's crust are oxygen and silicon. They make up 74.32% of the total and they are often combined in compounds called silicates. The other common elements – aluminium, iron, calcium, sodium, potassium and magnesium – are often combined in silicates and other minerals.

Oxygen
Silicon
Aluminium
Iron
Magnesium
Calcium
Sodium
Potassium

Conrad discontinuity
Basalt
Mohorovicic discontinuity
Sedimentary rocks
Granite

Sediments
Basaltic pillow-lavas
Basalt
Mantle

km
0
10
20
30
40
50
60
70

Continental crust

Oceanic crust

Above: The section through the Earth's crust shows that the heavy basaltic layer on the sea floor and underlying the continents is separated from the lighter granitic rocks of the continents by the Conrad discontinuity. The boundary between the crust and the dense mantle is the Mohorovicic discontinuity.

troyed in the Biblical Flood.
Lustre is a description of how light is reflected from a mineral. For example, adamantine lustre is bril-

Sir Charles Lyell

liant, like a diamond; vitreous is like glass; specular is like a mirror; and so on.
Lyell, Sir Charles (1797-1875) taught and developed tne theories of James Hutton, such as UNIFORMITARIANISM.

M **Magma** is liquid igneous rock. Its most familiar form is LAVA, which erupts from volcanoes. When magma solidifies, it forms IGNEOUS ROCK.
Magnetic field, Earth's. The Earth is like a giant magnet. The electricity which causes the magnetism is probably created by

Kilauea's crater in Hawaii

movements in the Earth's liquid outer core. As a result, the Earth has magnetic north and south poles, lying close to the geographic poles.

Mantle, Earth's. The mantle, between the core and the crust, is about 2,900 km thick. It is mostly solid and composed of dense

rocks, but beneath the crust, parts of it may be liquid.
Mesozoic era lasted between 225 and 63-64 million years ago. Mesozoic means 'middle life' and the era was the time when reptiles ruled the Earth.
Metamorphic rocks are rocks that have been changed by heat, pressure or chemical activity. For example, pockets of MAGMA give off much heat and cause contact metamorphism, whereby rocks around the hot magma are changed.
Minerals have definite chemical compositions. This

1 Talc (softest)	2 Gypsum	3 Calcite	4 Fluorite	5 Apatite	6 Orthoclase	7 Quartz	8 Topaz	9 Corundum	10 Diamond (hardest)

per cent of the Earth's crust is made up of eight elements: oxygen (46·6 per cent), silicon (27·7 per cent), aluminium (8·1 per cent), iron (5·0 per cent), calcium (3·6 per cent), sodium (2·8 per cent), potassium (2·6 per cent) and magnesium (2·1 per cent).

Minerals may be single, or native, elements or they may be compounds (combinations) of elements. Minerals are homogeneous – that is, each part of a mineral is the same as any other. Each also has a definite and unique composition and character. The largest group of minerals are called SILICATES. These are compounds of the two most common elements, oxygen and silicon, with much smaller amounts of other elements.

Two minerals with the same chemical make-up may have very different characteristics. For example, the element carbon can take the form of an extremely hard diamond or of soft graphite in a pencil. The difference in form is caused by the arrangement of the atoms in the mineral. Arrangements of atoms can also make a mineral brittle or flexible, heavy or light, colourless or

Above: In the Moh's scale of hardness, used to identify minerals, common minerals were given numbers from 1–10, in order of hardness. Gypsum (2) will scratch the softer talc (1), but gypsum is scratched by the harder calcite (3). Diamond (10) is the hardest mineral.

Below: The Canadian Shield is part of the Earth's crust that has, for millions of years, remained stable – that is, there has been no mountain building or other movement. This kind of flat, worn-down landscape also occurs in Western Australia, Brazil, the USSR and elsewhere.

colourful, clear or frosted, or shiny or dull.

The properties of minerals are transferred to the rocks they form. Some rocks are composed of only one mineral, but most contain two or more minerals. Most rocks are heterogeneous – that is, their composition is varied. For example, GRANITE contains feldspar and quartz, sometimes with smaller amounts of hornblende and mica. The proportions vary from sample to sample.

Rocks in the Earth's crust

Ever since the Earth was created, rocks have been forming and re-forming in the Earth's crust. There are three main types of rocks: IGNEOUS, SEDIMENTARY and METAMORPHIC.

Igneous rocks, named after *igneus*, the Latin word for 'fire', are formed from molten rock, or magma. When magma cools slowly beneath the Earth's surface, it forms rocks such as granite, which contain large crystals. These rocks are called intrusive igneous rocks. When magma emerges on the surface as lava, it cools quickly. It then forms fine-grained rocks, with small crys-

distinguishes them from rocks, which contain varying amounts of minerals. COAL, oil and natural gas were

Oolitic limestone pillar

formed from once-living material and are not, strictly, minerals.
Mohole project was a US government project to drill through the Earth's crust to the Mohorovicic DISCONTINUITY to obtain samples from the mantle below. But after costs rose to 20 times the original estimate, the project was abandoned in 1966.
Mohorovicic discontinuity, see DISCONTINUITY.

O Ordovician period lasted between about 530 and 440 million years ago. It was thought, until

recently, that the first vertebrates (armoured fishes) evolved in this period. But evidence published in 1978 suggests that they were in existence in the preceding Cambrian period.
Ore is a combination of valuable minerals and other rocks, which are removed after the ore has been mined.

P Palaeontology is the study of ancient life and conditions through FOSSILS.
Palaeozoic era lasted from about 570 until 225 million years ago. Palaeozoic means

'ancient life'.
Permian period lasted from 280 until 225 million years ago. It is named after the Perm district in the Ural mountain area of the USSR.
Pre-Cambrian is a term used for the time before the CAMBRIAN PERIOD, the first in the Palaeozoic era. It covers about 4,000 million years, more than four-fifths of the Earth's total age. Fossils are rare in Pre-Cambrian rocks, probably because most organisms lacked hard parts.

Q Quartz is a common mineral, found in many

IGNEOUS ROCKS. Fragments of quartz accumulate to form sandstones, which are SEDIMENTARY ROCKS.
Quaternary period covers the last 2 million years. It is divided into the Pleistocene and Holocene (Recent) epochs. The term Quaternary originates from the early days of geology when Earth history was divided into four periods: Primary, Secondary, Tertiary and Quaternary. The first two terms are no longer used, but TERTIARY and Quaternary have survived as names of periods in the Cenozoic era.

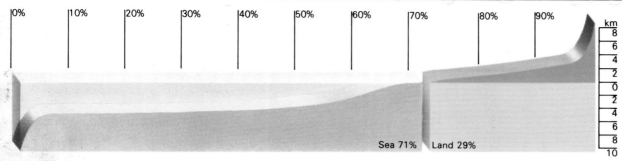

0% 10% 20% 30% 40% 50% 60% 70% 80% 90%

km
8
6
4
2
0
2
4
6
8
10

Sea 71%   Land 29%

Left: This diagram shows vertical heights and the proportions of the Earth's surface at various heights. It is called the hypsographic curve. The fairly level continental surfaces and the ocean floor are separated by the steep continental slope at the edge of the continental shelf.

tals. These extrusive igneous rocks are sometimes glassy, containing no crystals.

The rocks on the Earth's surface are constantly being attacked by rain, running water, wind, moving ice, solar heat, and chemicals dissolved in water. As a result, rocks are broken down into loose material or sediment. This sediment is often carried long distances by rivers, glaciers and so on. Finally, it comes to rest, usually in water, and accumulates in layers. These layers are compacted and chemically cemented together to form sedimentary rocks. Some contain large pebbles and are called conglomerates. Others, such as SANDSTONES, are composed of sand grains. Tiny particles of silt and clay form SHALES. Some sedimentary rocks consist of the remains of once-living organisms, while others form from chemicals dissolved in water.

Metamorphic rocks are rocks that have been changed by heat, pressure or chemical action.

Below: Rocks form in several ways. Granite and basalt are igneous rocks, formed from once molten magma. Granite solidified slowly underground, while basalt hardened quickly on the surface. Slate is a metamorphic rock, formed when shale, a sedimentary rock, is heated by magma and compacted under pressure. Conglomerates are sedimentary rocks composed of coarse rock fragments, which are cemented together. Oolitic limestone, another sedimentary rock, is formed in deep seas by the accumulation of tiny balls of calcium carbonate.

For example, hot magma bakes the surrounding underground rocks, causing limestone, for instance, to change into marble. Other rocks undergo incredible pressures caused by Earth movements, such as rock folding. The structure of such rocks is thus altered. For example, soft shale is pressed into hard slate. Chemical changes occur when hot steam or water, containing dissolved chemicals, reacts with rocks.

Because of movements such as CONTINENTAL DRIFT *(see page 76)*, rocks may be forced back into the Earth and remelted. This new magma may again rise to the surface through volcanoes.

Sedimentary rocks cover about 75 per cent of the Earth's land area, but form only 5 per cent of the top 16 kilometres of the Earth's crust. The rest are either igneous or metamorphic. The three types occur together where igneous rock intrudes into sedimentary rock. Metamorphic rock forms in the contact zone.

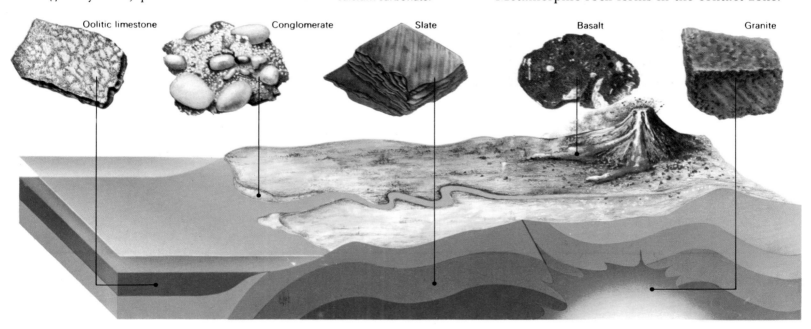

Oolitic limestone   Conglomerate   Slate   Basalt   Granite

Land's End, England

R **Rocks** do not have a definite chemical composition, because they are composed of differing amounts of MINERALS. There are three kinds of rocks: IGNEOUS, METAMORPHIC and SEDIMENTARY.

S **Sedimentary rocks** include *clastic rocks*, formed from fragments of other rocks which accumulate, in layers, usually in water; *organic rocks*, such as coal, formed from the remains of once-living material; and chemically-formed rocks, such as evaporites, which form when water is evaporated from a solution. Clastic rocks are classified according to the size of the grains they contain. For example, *rudaceous rocks*, such as breccias and conglomerates, are coarse-grained, containing particles more than 2 mm across.

Rock strata on the Northumberland coast, England

Arenaceous rocks, such as sandstone, contain particles between 0.06 and 2 mm across. *Argillaceous rocks*, such as siltstone, have grains between 0.0025 and 0.06 mm across. Clays have even finer particles.
Silicates are the most important group of minerals, making up 95% of the minerals in the Earth's crust.
Silurian period lasted between 440 and 410 million years ago. It was named after the Silures, an ancient Welsh tribe, who once lived in the area where Silurian rocks were first studied.

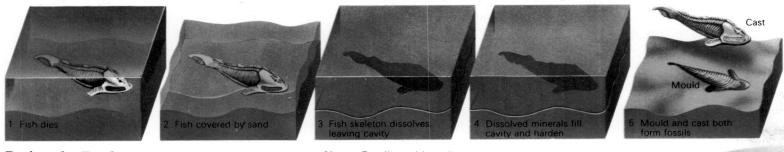

1 Fish dies

2 Fish covered by sand

3 Fish skeleton dissolves, leaving cavity

4 Dissolved minerals fill cavity and harden

5 Mould and cast both form fossils

Cast

Mould

Dating the Earth

The Earth has been in existence for 4,600 to 5,000 million years. Such an age is difficult to imagine. If the Earth's entire history is compared with a record with 23 minutes of playing time on each side, then Man would appear just two and a half seconds from the end. The last 2,000 years would not even amount to a crackle.

The study of fossils made scientists realize that the Earth must be extremely old and it has helped geologists to place events in the Earth's history in order. Exact dating was, however, impossible until the discovery of radioactivity.

Radioactive DATING is based on the fact that some materials found in rocks are unstable. They break down, or decay, by emitting radioactive

Above: Fossil moulds and casts were formed when a dead creature, such as an armoured fish, was buried quickly. The fish rotted and was dissolved away, leaving a hollow mould in the rock. Later, the mould was often filled by minerals, creating a fossil cast of the original fish.

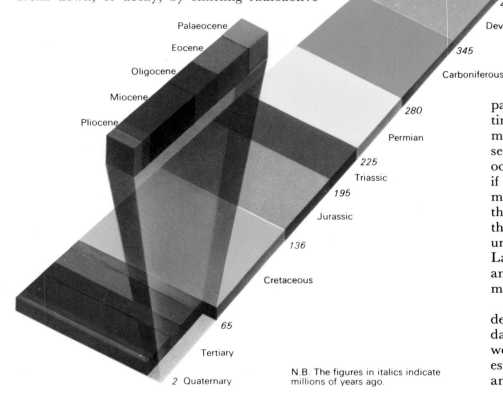

Palaeocene
Eocene
Oligocene
Miocene
Pliocene

Pre-Cambrian
570
Cambrian
530
Ordovician
440
Silurian
410
Devonian
345
Carboniferous
280
Permian
225
Triassic
195
Jurassic
136
Cretaceous
65
Tertiary
2 Quaternary

N.B. The figures in italics indicate millions of years ago.

Left: The last 570 million years of the geological time scale are divided into 3 eras: the Palaeozoic (570–225 million years ago); the Mesozoic (225–65); and the Cenozoic (the last 65 million years). Each era is divided into periods and the Cenozoic periods are divided into epochs.

particles, until a stable particle is formed. The time taken for radioactive decay depends on the material. It ranges from a few thousandths of a second to millions of years, but decay always occurs at a constant rate. Scientists realized that, if they measured the amount of radioactive material in the stable end product, they could then establish the time taken for the decay. This then would be the age of the material. At first, uranium, which decays into lead, was used. Later, rubidium, which decays into strontium, and potassium (to argon) were also used to measure long periods of time.

One radioactive substance, called CARBON-14, decays in less than 100,000 years. It is used to date fairly recent objects, such as boats and weapons. By radioactive dating, geologists have established dates for the Earth's eras, periods and epochs.

Colorado River gorge

During the Silurian period, the first land plants appeared.

Smith, William (1769-1839) was a surveyor who collected and studied FOSSILS while he was working on canal building in southern England. He later mapped the rock strata and published a geological map of southern England in 1815.

Stratigraphy is the study of the types and ages of layered, or stratified, rocks and the mapping of them.

T **Tertiary period** lasted between 63-64 and

about 2 million years ago. It is divided into 5 epochs and was the time when mammals developed into many forms. For the origin of the term Tertiary, *see* QUATERNARY PERIOD.

Triassic period lasted from 225 to 190-195 million years ago. It was so-named because the rocks of this period in Germany were found to be divided into three distinct layers. The first dinosaurs appeared in this period.

U **Uniformitarianism** is a theory put forward by

James HUTTON suggesting that whatever processes are operating now on the Earth also operated in the past. Hence, for example, by studying how sediments accumulate in water, we can understand how SEDIMENTARY ROCKS were formed.

W **Werner,** Abraham Gottlob (1749-1817), a leading German mineralogist, believed that most rocks were formed from chemicals dissolved in a vast ocean. He also thought that volcanoes were coal seams burning underground. His followers

Abraham Werner

were called Neptunists, after the god of the sea, but his theories were discredited by the Plutonists (*see* HUTTON, JAMES).

The land beneath our feet is moving as crustal plates are propelled around by forces
within the Earth. Oceans are widening and mountains are rising. Linked with plate
movements are two terrifying phenomena – earthquakes and volcanoes.

The Restless Earth

The Earth's surface may seem solid and immovable enough, but we now know that the Earth's crust is cracked and parts of it are moving around slowly, like rafts on the mantle below.

The German meteorologist Alfred WEGENER first suggested the theory of CONTINENTAL DRIFT in 1912. Like earlier scholars, he drew attention to the similarity between the shapes of the Americas and Europe and Africa and claimed that they were once joined together. To support his theory, he collected evidence of similarities between fossils, rocks and rock structures found on both sides of the Atlantic Ocean.

Most geologists rejected the theory, particularly because Wegener could not explain how the continents might have moved. However, in the 1960s, other scientists produced the plate tectonics theory, which is now generally accepted.

Continents on the move

The plate tectonics theory suggests that the Earth's crust is broken like a cracked egg shell. Various pieces, called PLATES, are floating on the denser mantle below. The Earth's main surface features – the continents and the oceans – rest on top of the plates, which are moved by convection currents in the mantle. These convection currents in fluid rock are much like those that occur when you heat a pan of water. Water particles near the heat rise to the surface and then spread horizontally across it. Finally, the particles cool and sink again to the bottom of the pan. In the mantle, horizontal spreading of fluid magma beneath the crust is one force which can move the plates.

Another dynamic force operates beneath the thin oceanic crust. Along enormous underwater mountain ranges (even higher than Mount Everest in places) called MID-OCEANIC RIDGES, rising magma breaks through the crust along long cracks. This magma forms new rock along

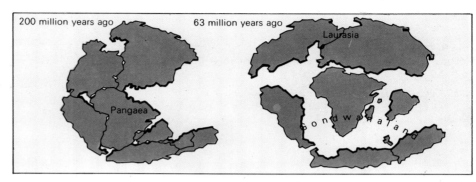

Above: About 200 million years ago, the original continent of Pangaea began to split into Laurasia and Gondwanaland. Then, these landmasses also split up and the separate parts drifted to their present positions.
Below: The oceans are widening as new rock is added to the mid-oceanic ridges. The diagram shows how ocean-spreading is pushing plates 2 and 3 against each other. The edge of plate 2 is forced under plate 3 in a subduction zone. This zone has a deep oceanic trench. When plates collide, rocks are folded up into mountain chains.

the ridges and so the ocean floor is widened. This process is called SEA-FLOOR SPREADING. Evidence of sea-floor spreading has been found in ICELAND, an island outcrop of the Atlantic mid-oceanic ridge, which is also slowly getting wider.

Scientists believe that, around 200 million years ago, all the continents were joined together in a super-continent, called PANGAEA. This continent first split apart into two smaller continents, GONDWANALAND and LAURASIA, which also broke up in the last 180 million years. Gondwanaland split into South America, Africa, Antarctica, Australia and India. Laurasia was divided into North America, Europe and most of Asia. India was joined to Asia only in recent geological time (*see page 78*).

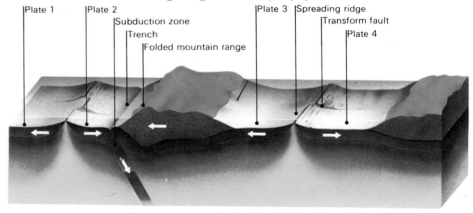

Plate 1 | Plate 2 | Subduction zone | Trench | Folded mountain range | Plate 3 | Spreading ridge | Transform fault | Plate 4

Collision course

Like giant conveyor belts, the ocean floors are spreading by about two centimetres a year. However, because the Earth is not getting any bigger, the moving plates are bound to collide.

When plates collide, several things can happen. First, the edges of two plates may slide past each other in a series of jerks, causing earthquakes. This occurs along the edges of the North American and Pacific plates. Second, one plate may be forced down below another. This movement also causes earthquakes and the descending plate is melted, creating magma which rises through nearby volcanoes. These phenomena occur around Japan, where the advancing edge of the Pacific plate is being destroyed as it is forced down under the Eurasian plate. Third, like the Japanese islands themselves, plates may collide in such a way that the edges are crumpled and pushed upwards into mountain chains or ISLAND ARCS. The ANDES were formed in this way by the collision of the Nazca and South American plates.

Evidence from the sea-floor

The study of the ocean floor in the 1950s provided much evidence to support the idea of continental drift. On each side of the mid-oceanic ridges, samples of rock were found to contain iron minerals which were lined up in the Earth's magnetic field, much like a mass of tiny magnets. When plotted on charts, the alignment of these minerals appeared as patterns of magnetic 'stripes'. The pattern on one side of the ridge is a mirror image of the other side. This suggests that, when the magma rose to the surface and solidified, it was divided evenly and the two halves moved in opposite directions.

Another feature of the sea-floor is that, geologically-speaking, it is extremely young. Continental rocks more than 3,000 million years old have been found, but no part of the ocean floor is more than 200 million years old. This supports the theory that the sea-floor is being constantly destroyed and renewed.

When the edges of the shallow continental shelves, the true edges of the continents, were mapped, the similarities between the shapes of continents, noticed by Wegener and others, were seen to be even closer. The fit was so accurate that it could not be coincidence.

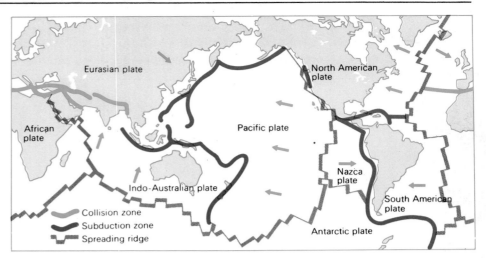

Above: The map shows the main plates into which the Earth's crust is divided. Around the plate edges there are volcanoes, and earthquakes occur when the plates move.

Below: A space photograph of southern Italy shows the Bay of Naples and the Gulf of Taranto, near Italy's heel. Volcanoes in south-western Italy are associated with a nearby plate edge.

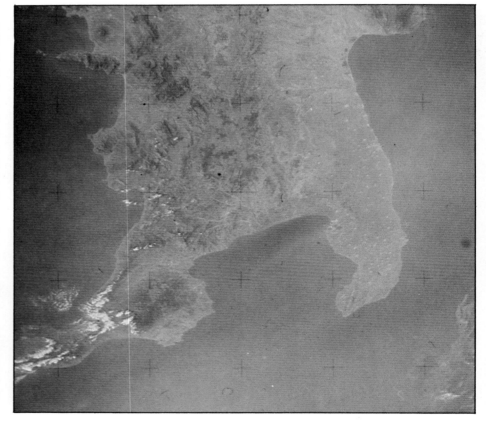

---

rock (nearly always granite). They often cover thousands of square kilometres, stretching along mountain building belts, with which they are associated. Some of the world's granite areas, such as in Cornwall, England, are exposed batholiths. Small batholiths are called laccoliths.
Ben Nevis, at 1,343 metres above sea level, is the highest mountain in the British Isles.
Block mountains are blocks of land raised up between faults. They are similar to HORSTS.

C Caldera is a large volcanic crater. It may be a combination of several small craters or it may be formed when a volcano explodes

Volcanic caldera in the Azores

and its vent collapses. Calderas often contain lakes.
Columnar basalt occurs when basaltic lava cools and shrinks into regular, 6-sided columns. Examples occur at the Giant's Causeway, Northern Ireland, and at Fingal's Cave, on the Scottish island of Staffa.
Continental drift is the theory which states that the continents were once joined together, but they have drifted apart in the last 180 million years (*see pages 75-77*).
Convection currents in the Earth's mantle move the plates which support the drifting continents.
Craters are depressions in the ground which look circular from the air. The two

main types are explosion craters, caused by volcanic action, and impact craters, caused by meteorites.

D Dyke is a steeply-inclined sheet of igneous rock that cuts across the beds of existing rocks. They often occur in swarms and may cover hundreds of kilometres. The term dyke is also used for sea walls and ditches.

E Earthquakes may be caused by volcanic eruptions, landslides or even nuclear explosions. But

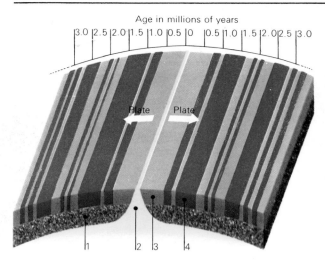

Age in millions of years
3.0 2.5 2.0 1.5 1.0 0.5 0   0.5 1.0 1.5 2.0 2.5 3.0

Plate   Plate

1   2 3   4

Left: As magma rises along the mid-oceanic ridges (2) from the mantle (1), iron minerals are lined up with the Earth's magnetic field. Periodically, the magnetism is reversed. On magnetic charts of the rocks near the ridges, there is a zebra-like pattern of normally-magnetized rocks (3) and reverse-magnetized rocks (4). The pattern is the same on both sides of the ridges, because the new rock is divided evenly and moved in opposite directions.
Right: Coal seams in Antarctica were formed when the land was in the tropics.

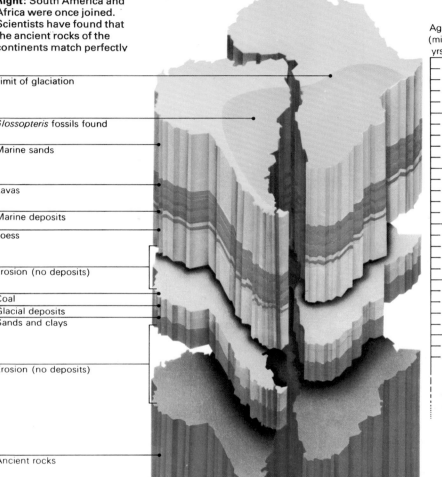

Evidence from other sources

In the late Carboniferous and early Permian periods, ice covered much of South America, Africa, Australia, India and Antarctica. Geologists have traced the paths of the great ice sheets of this Ice Age from marks on the rocks and from deposits of ice-borne material. The movements of the ice suggest that the land areas were joined together during that Ice Age.

Fossil evidence also supports the view that an unbroken landmass existed at some distant period of geological time. There is no other way to explain the broad sweep of continuous and similar plant growth. Prehistoric animals also provide evidence. Fossils of *Mesosaurus,* a reptile which lived about 300 million years ago, occur in South America and Africa. Unless it swam the Atlantic, in which case it would have spread much more widely, South America must have been joined to Africa at that time.

Further evidence comes from rock types, volcanic flows and rock structures on both sides of the Atlantic. For example, the mountain chains of Norway and Scotland are, structurally, a continuation of those on the east coast of North America.

The theory of plate tectonics explains many mysteries. For instance, the edge of the Pacific Ocean is often called the 'ring of fire', because it is an earthquake and volcanic zone. This zone coincides with the edge of the Pacific plate. In the east, around California, two plates are sliding past each other, causing earthquakes but not volcanic activity. However, in the west around

Right: South America and Africa were once joined. Scientists have found that the ancient rocks of the continents match perfectly

Limit of glaciation

Glossopteris fossils found

Marine sands

Lavas

Marine deposits

Loess

Erosion (no deposits)

Coal

Glacial deposits

Sands and clays

Erosion (no deposits)

Ancient rocks

Age (mill. yrs.)
0
100
200
300
400
500

---

most of them are caused by sudden movements along FAULTS. They may occur anywhere, but the most severe usually occur on or near the edges of PLATES.
Epicentre is the point on the Earth's surface directly above the focus (point of origin) of an earthquake.
Exhalation is the expulsion of volcanic gases and hot steam at the Earth's surface.

F **Fault** is a fracture, or crack, in rocks along which there has been movement. The rocks on either side of a fault are often

crushed into a powder, called cataclastic rocks. Faults are ideal routes for fluids to rise to the surface and deposit minerals.

Earthquake damage in Anchorage, Alaska (1958)

Fire fountains are jets of burning gas and liquid lava which spurt from a volcano. They are always associated with basaltic lava and may

reach heights of 400 metres. Several fire fountains joined together are called fire curtains.
Folds are bent layers of rock. They are caused by lateral (sideways) pressure and they are especially common in fold mountain chains. The simplest types are anticlines (upfolds) and synclines (downfolds), but they also occur in other shapes and a great variety of sizes.
Fumarole is a hot spring, giving out hot gases and steam. The Valley of Ten Thousand Smokes in Alaska

is named after its many fumaroles.

Lulworth Cove, Dorset

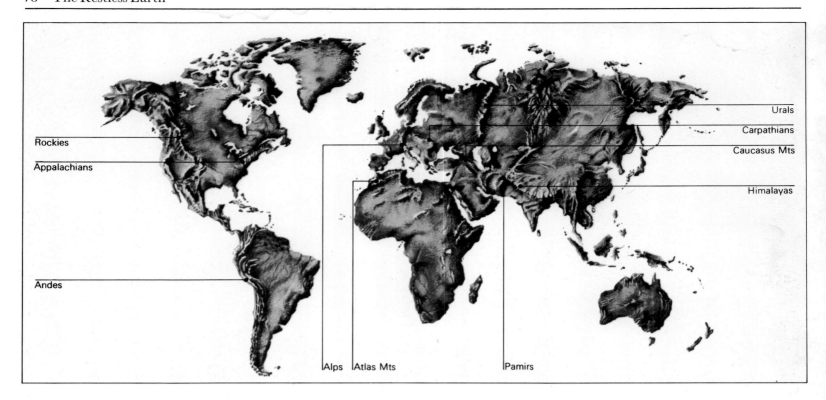

Rockies
Appalachians
Andes
Alps | Atlas Mts
Pamirs
Urals
Carpathians
Caucasus Mts
Himalayas

Japan, both earthquakes and volcanic action occur. The plate tectonics theory explains such differences and also provides clues about the inner workings of the Earth.

Mountains

Mountains have always intrigued man and some, such as MT. FUJI in Japan and Olympus in Greece, are or were regarded as sacred. Besides their scenic grandeur, mountains provide everything from water and hydro-electric power for industry to a challenge to intrepid climbers. They also provide geologists with clues about the Earth's deepest secrets. There are three main kinds of mountains: FOLD mountains; BLOCK MOUNTAINS (*see page 80*); and VOLCANOES.

The regular patterns produced by fold mountain chains stand out on maps, because they are formed in long, narrow belts — areas that have undergone periods of great movement and stress.

The birth of a mountain range

Fold mountains are formed when plates collide. For example, the great peaks of the HIMALAYAS, containing the world's eight highest mountains, were raised up when a plate, bearing what is now

Above: The map shows the world's chief folded mountain ranges. These ranges were squeezed upwards by lateral (sideways) pressure, as plates in the crust were pushed against each other. Some ranges, such as the recently-formed Alps and Himalayas, are probably still rising.

India, broke away from Gondwanaland and moved steadily northwards until it was forced against the Asian continent. From about 53 million years ago, the sedimentary rocks between the colliding plates were thrown upwards and wrinkled like a giant tablecloth to form the Himalayas. Finally, India became welded on to Asia. The ALPS were formed in a similar way, about 26 million years ago, when a plate supporting what is now Italy was rammed against Europe.

Mountains have exciting pasts and interesting roots. The average thickness of the crust under continents is only 35 to 40 kilometres. But, under mountain chains, the crust may be as much as 70 kilometres thick. Why is there such a difference? We know that sediment collects by the millions of tonnes on the continental shelves and in OCEAN TRENCHES, the deepest parts of the oceans. These ocean trenches form where an ocean plate is forced beneath a continental plate. While this is happening, volcanic action usually begins and volcanic rocks are also injected into the area. Eventually, the ocean closes, scraping up the sediment, crumpling the rocks and pushing them upwards. Because this thick mass of sediment is

---

G **Geochemistry** is the study of the chemical composition of the Earth and the distribution of elements within it. The geochemical cycle is the name for the way in which igneous rocks form from magma. The igneous rocks are then broken down to form sedimentary rocks or converted into metamorphic rocks. Finally all the rocks are buried and remelted to start the cycle all over again.
Geosynclines are long hollows in the Earth's crust where great thicknesses of sediment accumulate. The weight of the sediment de-

presses the hollows even more. Finally, the mass is uplifted and deformed to form fold mountains.
Geyser is an eruption, sometimes violent, of superheated water and steam from underground sources. It is caused by superheating, or possibly by the action of gases in the water, which act in the same way as gassy liquids explode from a bottle when you open it. Geyser regions include Iceland, New Zealand and the Yellowstone National Park, Wyoming, in the USA. The record erup-

Lone star geyser, USA

tion of 457 metres occurred at Waimangu Geyser in New Zealand in 1904. The highest active geyser is now Steamboat Geyser in Yellowstone National Park. It reaches heights of 76 to 115 metres, at intervals between 5 days and 10 months.
Gondwanaland was a continent formed when PANGAEA broke up 200 million years ago. In the last 180 million years, Gondwanaland has split into South America, Africa, Australasia, India (which joined with Asia), and Antarctica.
Graben is a trough-like de-

pression formed when land slips down between two parallel FAULTS. *See also* RIFT VALLEY.

H **Himalayas** are a great fold mountain range in Asia, formed by the collision of Asia and India (*see page 78*). It contains 96 of the world's 109 peaks that are more than 7,315 metres above sea level. The chain is about 2,500 km long. The highest peaks are: MT. EVEREST in Nepal-Tibet (8,848 metres); K2 (Mt. Godwin-Austen) in Kashmir-Sinkiang (8,610 metres); Kanchenjun-

lighter than the denser material below, it 'floats'. Through a process called ISOSTASY, the crumpled, deep mass of rock gradually rises until it reaches a point of equilibrium. This explains how high mountain chains are formed along the original edges of continents, and why fossil remains of sea animals are found near the tops of such peaks as MT. EVEREST.

New and old mountains

Even as a mountain chain is rising upwards, the forces of erosion, such as weathering, running water and moving ice, start to wear it down. Eventually, over millions of years, high, jagged peaks are reduced to flat plains.

Above: The Sun rises behind Annapurna III, one of the great Himalayan peaks of Nepal. The Annapurna glacier is in the foreground.
Above right: The Matterhorn, on the Swiss-Italian border, is one of the most popular Alpine peaks among mountaineers.
Below: The 10 highest mountains of the world are in Asia. Eight are in Nepal and the other 2 are in Jammu and Kashmir.

The world's highest fold mountains are, therefore, comparatively young. The Alps and the Himalayas are the youngest and they are probably still rising. In places, the movement is accompanied by violent earthquakes and volcanic action. The ROCKY MOUNTAINS and the Andes chain in the western part of the Americas are also relatively young and active. They contrast with the eroded Appalachians in the east, which have no earthquakes or volcanoes. The Appalachians are also many kilometres from any plate edge. The Scottish Highlands and the Russian URALS are older fold mountains.

Large tracts of flat land, on the other hand, are partly the roots of mountains which were eroded

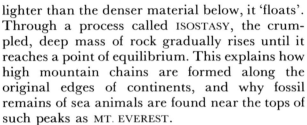

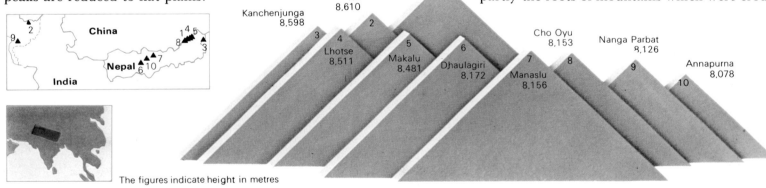

The figures indicate height in metres

ga in Nepal-Sikkim (8,598 metres); Makalu I in Tibet-Nepal (8,481 metres); Dhaulagiri in Nepal (8,221 metres) ; Nanga Parbat in Kashmir (8,126 metres); and Annapurna in Nepal (8,078 metres). However, the Himalayas cannot compare in size with the world's greatest mountain chain — the Mid-Atlantic Ridge in the Atlantic Ocean which is 16,000 km long and 800 km wide.
Horst is a ridge-like block of land raised up between two parallel FAULTS. An example is the Harz Mountains.

Hydrothermal processes are the action of heat and water in igneous activity. Superheated water can alter and break down rocks and

Geothermal power station, New Zealand

minerals. It can also carry elements in solution over long distances before depositing them. Temperatures of the superheated

water range between 100°C and 500°C.

Iceland, in the Atlantic Ocean, is one of the few places where the Mid-Atlantic Ridge appears above sea level. It contains active volcanoes and often experiences earthquakes. It is virtually being pulled apart by SEA-FLOOR SPREADING. In 1963, just off the coast of Iceland, a volcano appeared to form the new island of Surtsey.
Intrusions occur when bodies of igneous rock are forced into existing ones.

Island arc, also called a festoon, is a chain of islands found near the edges of plates, where one PLATE is forced beneath another, such as off the coast of Japan. They are associated with earthquakes, volcanoes and deep OCEAN TRENCHES.
Isostasy is the Earth's elastic quality. Like a sponge, the crust springs back into shape whenever the pressure on it has been relieved. For example, areas of the continents are often depressed by heavy ice sheets. When the ice melts, the crust rises back to its original

long ago. These SHIELD or platform areas cover large parts of Canada, Australia and Siberia. These shields are ancient Pre-Cambrian landmasses, which have been in existence almost since the start of geological time. They are the complete opposite of the high peaks of young mountain ranges, which have not been around long enough to be worn down. Most continents combine these contrasting types of landscape.

Folding and faulting

With continental movements and violent mountain building, it is hardly surprising that the Earth shakes and rocks bend and break.

Although most rocks were originally laid down in flat layers, few remain that way for long. The twisted and fractured shapes often exposed in cliff faces and road cuttings clearly show how even the hardest rocks can be deformed. Rocks can be squeezed and shaped as if they were made of putty. In fact, some rocks 'flow' when under pressure. For example, salt may flow to the

Below: The diagram shows structural features caused by movements in the Earth's crust. Rocks are often broken to form large cracks or faults. There are various kinds of fault, depending on the direction in which rocks move along them. When a block of land sinks between 2 faults, a graben or rift valley is formed. When a block of land is pushed upwards between 2 faults, a horst or block mountain is formed. Sometimes, blocks of land move laterally (sideways) along a fault. Thrust faults occur when a fold is sheared and the top part is pushed over the bottom part.

surface in domes. It is sometimes powerful enough to bend the overlying rocks.

Folds can be several kilometres long. The two main parts of a fold are the anticline (the upfold or arch-like part) and the syncline (the downfold). Folds can be extremely complex and may be refolded many times during the formation of a mountain range.

Most rocks will take only so much bending before they crack and break. Cracking forms vertical joints in rocks. But, when each side of the crack begins to move, a FAULT is created. Faults vary in size from a few metres to others, called geofractures, which can extend across a continent. Often, one side of the fault drops relative to the other. Therefore, some parts of the land are raised up high above other parts. They are separated by a fault plane.

If two parallel fault planes are formed, the centre may drop down to form a GRABEN, or a long RIFT VALLEY. Alternatively, a block may rise up to form a HORST, a high, straight-sided ridge,

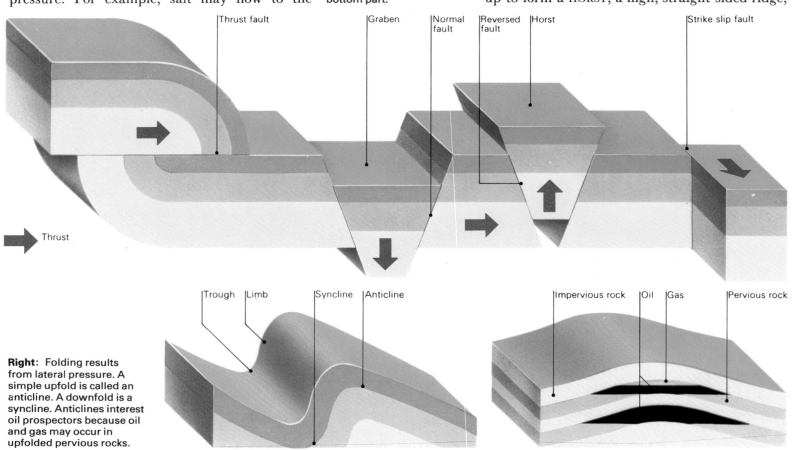

Right: Folding results from lateral pressure. A simple upfold is called an anticline. A downfold is a syncline. Anticlines interest oil prospectors because oil and gas may occur in upfolded pervious rocks.

level. Isostasy also occurs when mountains are worn down, because their deep roots compensate by pushing upwards to restore the balance.

J Joint is a fracture in a rock where, unlike a FAULT, there is no movement between the two sides. They are formed in several ways. For example, shrinkage joints are caused by the cooling and contraction of igneous rocks.

K Krakatoa, a volcanic island in Indonesia, ex-

ploded on August 27, 1883. Only 3 small parts of the rim were left, as islands, after the explosion. A cloud of ash rose 80 metres into the air

Geological fault in Campsie Glen, Scotland

and a 36-metre-high TSUNAMI battered the coasts of nearby Java and Sumatra, killing 36,000 people. The ash caused vivid sunsets.

L Laccolith, see BATHOLITH.
Lahars are mud flows associated with volcanic eruptions. They may be caused by heavy rain which sweeps loose ash or other debris down the side of a volcano, by the melting of snow and ice, or by the destruction of a CRATER lake.
Lassen Peak, which is 3,187 metres high, is in the Cascade Range of northern California. It is the only active volcano in the United States, last erupting in 1921.
Laurasia was an ancient continent formed when PANGAEA broke up about 200

million years ago. It later split into North America and Eurasia. Other landmasses formed GONDWANALAND.

M Mauna Kea is the world's highest mountain if measured from its base on the sea bed; it is 10,203 metres high. However, only 4,205 metres are above sea level. It is on the island of Hawaii.
Mauna Loa is an active SHIELD VOLCANO in Hawaii, 4,167 metres above sea level. A great crater, called Kilauea, on the volcano's flank, gives out lava as well

Below: The map shows how most earthquakes and volcanic activity are associated with the regions around plate edges in the Earth's crust.

Right: Some earthquakes occur when 2 plates move alongside each other. The plates often become jammed and the pressure mounts until it is released in a sudden, violent jerk. The focus is the point of origin of the earthquake. The epicentre is the spot on the Earth's surface above the focus.

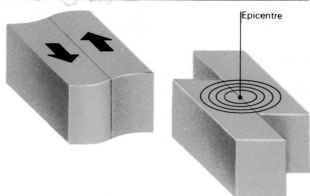
Epicentre

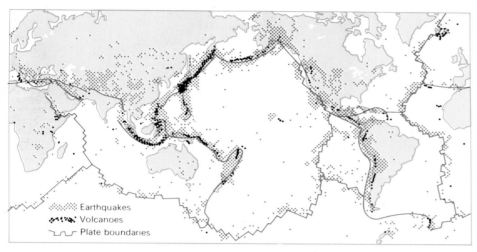
- ░░░ Earthquakes
- ◆◆◆ Volcanoes
- ⌐_⌐ Plate boundaries

such as the Vosges mountains in France. Large horsts, such as the North American Sierra Nevada mountains and the Ruwenzori range in East Africa, are called block mountains. The best-known rift valleys are those of the Rhine and Jordan rivers, and the East African lakes area.

Often the movement along faults takes place gradually over millions of years. However, if there is no lubrication, the rocks may stick together. When this happens, pressure builds up and it may finally be released in a sudden, violent jerk, causing an earthquake.

Below: This scene of devastation was photographed in Peru in 1970 after an earthquake had caused an avalanche on the slopes of Nevados Huascarán, Peru's highest peak. Snow and ice crashed down at about 480 km/h, breaking off rocks and boulders. The debris struck the village of Yungay with tremendous force. Only the tops of telegraph poles testify to the former existence of the village.

Earthquakes

When rocks move suddenly along a fault, the energy that has built up is released. This energy may cause catastrophes on the Earth's surface. The point of origin of an earthquake, which may be far underground, is called the focus. The point on the surface directly above the focus is called the epicentre.

The most destructive earthquakes have a shallow focus, less than 70 kilometres from the surface. The effects at the surface vary from sharp vibrations and shaking, to the land rolling like a gently ebbing sea.

The intensity of earthquakes may be measured in several ways. The Richter Scale measures the magnitude of the ground movements from 0 to 9. Because the scale is logarithmic, each successive number on the scale represents an increase in magnitude of ten times. A two-point magnitude on the scale represents an earthquake that is barely noticeable, while a seven-point magnitude earthquake is severe. The highest magnitude yet recorded is 8·9 on the Richter Scale.

as Mauna Loa itself. The thin, basaltic lava can flow up to 80 km from the cone.
Mid-oceanic ridges are long, narrow mountain ranges which rise from the sea-floor. Scientists have discovered that magma is welling up and adding new crustal rock along these ridges. Hence, they are PLATE boundaries, where SEA-FLOOR SPREADING is occurring.
Minoan civilization. The Minoans founded a major ancient civilization in the eastern Mediterranean, centred on Crete. Its rapid decline, about 3,500 years ago,

has been attributed to earthquakes and a violent volcanic eruption, which occurred on the nearby island of Santorini (now Thera), another Minoan centre, in about 1470 BC. Some people believe that the Santorini eruption gave rise to the legend of Atlantis.

San Pedro volcano on Lake Atitlan, Guatemala

Mount Cotopaxi, in Ecuador, South America, is one of the world's highest active volcanoes. It is 5,896 metres high and is south of Quito, Ecuador's capital. Even though it is only 80 km from the equator, snow covers the peak all the year round. As a result, LAHARS and floods, caused by lava melting the snow, are special dangers.
Mount Etna, in Sicily, is Europe's highest active volcano. It is 3,274 metres above sea level. More than 260 eruptions have been recorded since 700 BC.

Mount Everest is the world's highest mountain, towering 8,848 metres above sea level. Situated on the Nepal-Tibet border, it is

Miniature mud volcano

The Mercalli Scale measures the intensity of an earthquake from I (not felt by people even in the main area) to XII (when the damage is total). The Seismic Risk Scale measures the probability of damage or injury in certain areas of the world.

Effects of earthquakes

The longest recorded earthquake, which occurred in ALASKA in 1964, lasted seven minutes. This was the world's most severe recorded 'quake with a magnitude of 8·9 on the Richter Scale. However, most earthquakes last only a few seconds, but the damage done in that time can destroy entire cities. The possibility of damage increases if the soil and rocks are loose or soft. Soft ground moves and shifts more readily than hard rock. Soft clays and sand turn into a fluid due to the vibrations. Houses may sink into the ground and coffins may float up!

Ground movements may also start avalanches. For example, during an earthquake in Peru in 1970, snow and ice from the country's highest peak, Nevados Huascarán, broke away like a giant snowball. It collected debris and hit the village of Yungay at 480 kilometres per hour.

Many earthquakes occur under the sea. Although little damage is done directly, the movement generates waves, called TSUNAMIS, in the oceans. These waves, sometimes incorrectly called tidal waves, travel as fast as 750 kilometres per hour and can stretch for over 800 kilometres. In the middle of the ocean, the wave is only about a metre high but, when it approaches land and shallow water, the wave height increases up to more than 60 metres. These terrible waves can smash coastal towns to pieces. The explosion at Krakatoa in 1883 caused tsunamis to hit the Java coast, killing 36,000 people.

Earthquake danger zones

Earthquakes are much more common in some parts of the world than others. Areas near or at the edge of the plates, which are separated by giant faults, are prime candidates for earthquakes. The rim of the Pacific Ocean, famous for its violent history, produces four out of every five earthquakes and the Mediterranean and Middle Eastern zones are other danger areas.

Despite scientists' growing knowledge and the fact that the Earth is being constantly monitored

Above: The area traversed by the Wellington fault, a crack in the Earth's crust in New Zealand, would appear to be ideal for road-building, until the fault moves! Despite warnings, building still continues in fault zones.

Right: The circles on the map are isoseismal lines (lines that link points of equal intensity) from 3 earthquake tracking stations in the USA. The epicentre of the earthquake (the point on the surface above the point of origin of the earthquake) is the spot where the 3 circles meet.

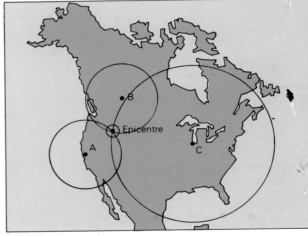

part of the HIMALAYAS. It was named after Sir George Everest (1790–1866), Surveyor-General of India. It was first climbed on May 29, 1953 by Sir Edmund Hillary and Sherpa Tenzing Norgay.
Mount Fuji, or Fujiyama, is a dormant volcano. At 3,777 metres above sea level, it is Japan's highest peak. It has been regarded as sacred since ancient times and is particularly holy to Shintoists. It last erupted in 1707.
Mount Kilimanjaro, a volcanic mountain on the Kenya-Tanzania border, is the highest mountain in

Africa. It is 5,895 metres above sea level. It is known locally as the 'Mountain of the Cold Devils'. Even though it is almost on the equator, parts of it are snow-capped all the year round.
Mount McKinley, in Alaska, is the highest peak in North America, being 6,194 metres above sea level.

N **Nappe** is a FOLD in rocks where the top part of the fold has been sheared away and pushed forward, sometimes for several kilometres, masking all the underlying rocks.

Nuées ardentes are 'glowing clouds' of hot gas, steam and VOLCANIC ASH, which occur during some volcanic eruptions, such as that of Mont Pelée in 1902 (*see page 86*).

Mount Egmont, New Zealand

Mont Pelée in 1902 (*see page 86*).

O **Ocean trenches** are the deepest parts of the

oceans. The deepest is the Marianas Trench in the Pacific Ocean, which is 11,033 km deep. Trenches occur where one PLATE is forced beneath another, thus forming earthquake zones.
Old Faithful is probably the most famous GEYSER in the world, sited in Yellowstone National Park, USA. It was so named because its eruptions, numbering 21 to 23 per day, have been consistently occurring since its discovery in 1870. The eruptions average about 40 metres high and discharge about 40,000 litres of water.

I. Measured only by seismographs

II. Tall buildings shake

III. Houses and high trucks are affected

IV. Pictures swing on walls

V. Glass shatters

VI. Leaves are shaken from trees

VII. Bells ring

VIII. Chimneys collapse

IX. Roads break up

with expensive and sensitive equipment, such as SEISMOGRAPHS, earthquakes are difficult to predict or control. Some areas have up to 600 tremors a day, but it is hard to forecast which, if any, will be severe. Precautions are taken in certain areas to minimize the danger. In earthquake zones, buildings are now specially designed to sway with the moving land. However, little can be done if the land shakes beneath a great city, such as San Francisco.

One place where research into earthquake forecasting is proceeding on a large scale is China, a land much hit by earthquake disasters. In February 1975, the city of Haicheng, which had a population of 100,000, was evacuated, because a severe earthquake was predicted. Two hours after the evacuation, the earthquake struck, but only three people were reported killed. Various methods are now being used, including the measurement of any unusual tilts in the ground, which may be caused by rock deformation below. Scientists also check the amounts of a radioactive gas, radon, which is dissolved in well water. They think that this gas, which is normally trapped in rock strata, is released when rocks crack open, prior to an earthquake.

Great earthquakes

The earthquakes that are

Many earthquake accounts contain twists of fate. The Lisbon earthquake occurred on November 1, 1755. This was All Saints Day and many people were in church when the earthquake struck. Many were buried alive, while others perished in the ensuing fires. Those who escaped ran to the harbour, only to be greeted by a giant tsunami, which drowned thousands.

The TOKYO-YOKOHAMA EARTHQUAKE of 1923, the world's most destructive in terms of property destroyed, also claimed many lives. It occurred at lunch-time, when many people were preparing meals on traditional open fires. After the fires were upset, the wooden buildings were soon ablaze and it was fire that did most of the damage and claimed most of the lives lost.

Fire also caused much devastation in the 1906 San Francisco earthquake. The city is built near the great SAN ANDREAS FAULT, a plate edge where the plates are moving alongside each other. But the plates do not move smoothly. Because their edges are jagged, they become jammed together until the pressure builds up to such an extent that, finally, the plates lurch along in a sudden, massive movement. This happened in 1906, when the plate edges near San Francisco moved horizontally by about 4·6 metres. The tremors caused by this violent movement led to the collapse of thousands of buildings. Gas pipes were snapped and electrical short-circuits caused fires to sweep through the shattered city. Experts, who have studied the San Andreas fault, have predicted that another major earthquake might occur at any time. Recently, scientists

golden fleece

orogeny, in the mid-Tertiary period, threw up the ALPS and the HIMALAYAS.

quakes because it is the edge of the moving Pacific

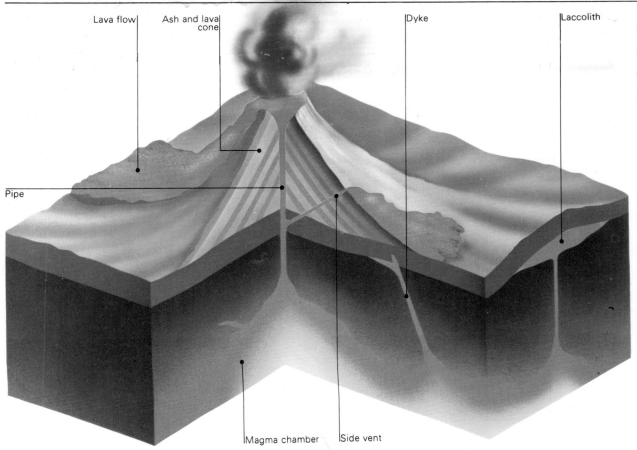

Lava flow

Ash and lava cone

Dyke

Laccolith

Pipe

Magma chamber

Side vent

Left: The diagram shows a section through a volcano and the molten magma chamber beneath it. The magma reaches the surface through a vertical pipe. But, occasionally, side vents occur and 'parasitic' cones develop on the volcano's flanks. Sloping sheets of magma, which solidify beneath the surface, are called dykes. Similar horizontal sheets of magma are called sills. Sometimes, large pockets of magma, called laccoliths, also solidify underground. They often push up the overlying rock layers. Large laccoliths are called batholiths. Some volcanoes erupt explosively and build up steep-sided ash cones; others emit streams of lava. But many volcanoes have composite cones, formed by layers of both ash and lava.

Below: Crater Lake occupies the crater of Mount Mazama, an inactive volcano in the Cascade Mountains of Oregon, in the USA. The lake covers about 50 square km and is about 610 metres deep. It formed slowly after the top of the volcano collapsed inwards.

have been investigating the possibility of relieving the tension along the fault. They think that it would be possible to bore a series of wells along the fault. If they pumped out underground water from two wells, they think that this would lock the fault at those points. If then, they pumped water into a third well between the two locked points, the water would lubricate the fault and cause a small earthquake. Working along the fault in this way would induce a series of small 'quakes, but probably prevent a major one.

Volcanoes

Volcanic eruptions are constant reminders of the energy stored beneath our feet, where nuclear reactions are generating tremendous heat and pressure. Volcanoes are the Earth's safety valves through which liquid rock (lava) and hot gases escape to relieve the pressure that builds up in the magma chamber and vents (openings). The greater the pressure beneath the vent, the greater the eruption will be.

smooth or rope-like surface. **Pangaea** is the ancient super-continent first post- | of the world, which were joined together in one solid landmass about 200 million | act fi

a sacred mountain and make annual pilgrimages to its *summit. It last erupted in 1707 and so it is regarded as a dormant volcano, which means it could erupt again.*

Vesuvius

Mont Pelée

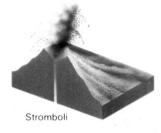

Stromboli

Vulcano

Mauna Loa

The close connection between volcanoes and earthquakes has been noted since the days of the ancient Greeks. The connection is now explicable in terms of the PLATE TECTONICS theory.

Dead or alive?

A volcano which has erupted in historic times is considered to be active. There are more than 500 active volcanoes in the world, including about 80 on the sea-floor. However, only about 25 active volcanoes erupt each year.

If evidence exists that a volcano has erupted in recent geological time, although no eruption has been recorded, then it is considered to be dormant. However, if there is no evidence of activity in the recent geological past, then it is pronounced extinct.

Even when a volcano is active, it may erupt only for a total of a few months in every million years. When it is dormant, it may be building up pressure. This pressure may be contained by a solidified lava plug which is plugging the vent.

Above: Volcanoes are often classified according to how they erupt. Vesuvius erupted explosively in AD 79. Much of the former mountain was destroyed and a cloud of gas, vapour and ash shot high into the air. No lava flowed, although most later eruptions were accompanied by lava flows. Mount Pelée in Martinique also erupted explosively in 1902, but it created a special feature, a nuée ardente (glowing cloud) which rolled downhill. Stromboli, an Italian island, erupts in moderate but repeated explosions. Lava is emitted during intense eruptions. Vulcano, another Italian island, emits viscous lava which hardens quickly, but it is shattered by explosive gases. Mauna Loa, Hawaii, is a 'quiet', non-explosive volcano. It has a low, shield-like shape.

Under a volcano

Beneath a volcano is a complex plumbing system, which connects the liquid magma to the surface. The link may be 60 kilometres or more in length. Deep down, tremendous pressures and temperatures maintain the magma as a gas-rich liquid.

When the magma starts to rise, the gases, including steam, carbon dioxide, sulphur dioxide, hydrogen and chlorine, may separate from the liquid rock and rise above it. The pressure accumulates under a lava plug in much the same way as pressure builds up when you shake a corked bottle of carbonated water. Eventually, the cork (or plug) is pushed out and the liquid and gas rush after it.

Explosive and quiet eruptions

If the magma in a volcano is highly charged with gases, the volcano usually erupts explosively. Magma with little gas causes quiet eruptions – that is, the volcano emits streams of molten lava,

Mount Robson, Canada

of volcanic
of superhe
lls down
lometres p
th fine vo
e volcan
s have ex
attered fr
t behind.

great vol
shattered the Mediterranean isla
(now Thera). This eruption is th
most powerful ever and abo
kilometres of rock were removed
recent explosion, which had about
power of Santorini, occurred at
island between Java and Sumatra
sound of the explosion was h
Springs, Australia, nearly 5,000 ki

One of the most famous ancien
that of VESUVIUS in AD 79. Ash fro
buried the town of Pompeii and

golden fleece

Left: Old Faithful is the best-known of the geysers in the Yellowstone National Park, Wyoming, in the USA. It lives up to its name, because it has erupted between 21 and 23 times per day ever since it was discovered in 1870. Each eruption throws about 40,000 litres of water some 40 metres in the air.
Right: Scientists believe that many geysers are powered by heat, such as that from magma. Some underground water is heated into steam. This forces the water above it upwards. After an eruption, the water drains back and is reheated until another eruption occurs. Some geysers may erupt because the water may become filled by gases. The volume of gases increases until the water bubbles upwards, in much the same way as carbonated water.

this continues, the Persian Gulf will eventually close up and a new fold mountain range will be formed from the sediments on the floor of the Gulf.
Seismographs are instruments to detect and measure the vibrations caused by earthquakes. The first, devised in 132 AD by the Chinese astronomer Chang Heng, contained finely balanced iron balls which fell when the ground shook. Modern instruments measure the intensity of earthquakes and distinguish them from nuclear explosions.

Seismological stations around the world provide information about the path of SHOCK WAVES through the Earth's interior.

A seismograph

Shensi province earthquake, in China on January 23, 1556, killed an estimated 830,000 people, the most ever in a single earthquake.

Shield volcanoes, in cross-section, resemble an upturned saucer. They are the product of quiet eruptions and thin runny lava.
Shields are large landmasses of Pre-Cambrian metamorphic and igneous rocks which have been undisturbed for much of the Earth's history. They are, therefore, the most ancient land areas. Examples are the Canadian and Baltic shields.
Shock waves are created by vibrations from earthquakes or explosions. There are 3 main types: *L-waves* are low frequency, long

wavelength
travel only
outer skin
earthqua
waves are
short wa
which trav
quake's fo
high fre
wavelength
travel throu
and liquid
The speed
depends o
the materia
through.
deduce the
the rocks i
terior fro

mud flow) engulfed nearby Herculaneum. More recently, in 1902, the eruption of Mt. Pelée on the West Indian island of Martinique was notable for its deadly nuée ardente, which killed 30,000 people in the city of St. Pierre. Parícutin, in Mexico, began life in 1943 when smoke appeared from a crack in a farmer's field. By 1952, when the eruptions ended, it stood 410 metres higher than the original level of the field.

While eruptions can destroy islands, they also account for new ones. One of the most recent is Surtsey, off Iceland, which first surfaced in 1963.

By-products of volcanic action

Volcanic gases and steam often rise to the surface in volcanic regions at temperatures up to 1000°C. The holes through which they emerge are called FUMAROLES. The gases often attack the rocks through which they pass and the rocks may be replaced with elements from the gases. Such elements form VEINS of valuable minerals.

Water circulating in the hot depths of the Earth heats up. When it comes to the surface, it forms hot springs, which are especially common in areas which have undergone recent volcanic activity. Some hot springs erupt jets of hot water and steam into the air. These are called GEYSERS. The record height of a geyser eruption is 457 metres by the Waimanga Geyser in New

Below left: The Mammoth Hot Springs are a beautiful tourist attraction in the Yellowstone National Park of Wyoming, in the USA. The water from hot springs is often highly charged with dissolved minerals. These minerals crystallize out of the water to form deposits. The spectacular terraces around the Mammoth Hot Springs consist of travertine, a kind of limestone.
Below right: The cliffs bordering the Isle of Staffa, an uninhabited island in the Inner Hebrides of Scotland, are lined by impressive basalt deposits. The basalt was formerly hot magma. But, as basaltic magma cools, it often shrinks and breaks into regular, 6-sided columns, resembling pillars.

Zealand. Hot springs and geysers often contain dissolved minerals, which crystallize out of the cooling water. Spectacular mineral deposits are, therefore, often found around them. Trees are coated with minerals and stone waterfalls are formed. Such deposits are called geyserite.

Rocks from volcanoes

Magma may come to the surface as volcanic ash, volcanic bombs and so on, but most of it as lava, which hardens to form igneous rocks. Basalt is a common rock formed from lava. It occurs widely on the sea-floor where it cools quickly into shapes that resemble masses of pillows. Hence, it is often called pillow lava. The magma which forms basalt is poor in the mineral silica. This means that it is runny and it often oozes quietly from cracks in the surface. The largest cracks are along the mid-oceanic ridges, where two plates are being pushed apart.

When basalt cools on land, the cooling is rapid and large crystals do not form. This fast cooling, therefore, gives basalt its fine-grained texture. Sometimes, basalt shrinks and breaks on cooling into regular six-sided columns, such as those at Giant's Causeway, Northern Ireland. Because of its liquid nature, basaltic lava can flow long distances over land. Parts of the continents are coated by great thicknesses of basalt.

Stromboli, off the coast of Sicily

Sill is an underground sheet of intrusive igneous rock that is parallel to the surface.
Stromboli, on the Lipari Islands off the coast of Sicily, is one of the few active volcanoes in Europe. The 926-metre-high cone last erupted in 1921.
Subduction zone is the area where a PLATE is forced down, or subducted, below another plate (usually near a continent's edge). It is usually associated with volcanic activity and earthquakes.

T Tokyo-Yokohama earthquake occurred in Japan on September 1, 1923. It claimed more than 140,000 lives, mainly because of the fires which destroyed more than 575,000 dwellings. The magnitude was 8·2 and it was the most destructive earthquake ever recorded (in terms of property).

Tsunamis, or seismic sea waves, are waves usually created by earthquakes and/or volcanic eruptions under the sea. In the open sea, the waves are low but they travel at 150–200 km/h. Near coasts, they increase in height to over 60 metres. One tsunami created during the Japanese earthquake of 1933 took 10 hours to cross the Pacific Ocean and reach San Francisco.

U Ural Mountains, in the USSR, average 1,800 metres in height. Running north-south for over 2,000 km, the range was formed during the Hercynian OROGENY, by the welding together of 2 separate landmasses. The highest peak is Narodnaya, at 1,894 metres.

V Vein is a sheet of minerals injected into a joint, fissure or gap in rocks. Most veins are filled by minerals deposited by gases or liquids. They may also be sedimentary in origin. Veins are important sources of valuable minerals.
Vesuvius, a volcano which overlooks the Bay of Naples in Italy, erupted explosively

Maximum height
reached by 1945

Size of cone after 24 hours

metres
500
400
300
200
100
0

1
2
Parícutin
3 km

The amount of lava that has spread over some areas is enormous. The Deccan plateau, in southern India, is covered by enough basalt to fill a vast cube with 700,000-kilometre-long sides. Basalt also forms volcanoes. The lava which flows from the quiet volcanoes of Hawaii solidifies into basalt. It flows great distances and, as a result, the so-called SHIELD VOLCANOES of Hawaii are shaped like an upturned saucer.

Other rocks from magma

Magma that does not reach the surface through volcanoes or cracks in the ground may solidify underground. Such magma cools slowly and so there is time for crystals to develop. One of the most common of these coarse-grained rocks is granite, which often forms in huge chambers called BATHOLITHS. Inclined sheets of magma are called DYKES and horizontal sheets are called SILLS. Granite and many other lavas, which are rich in silica, are extremely sticky and do not flow easily like basalt. Instead of pouring quietly from the vent of a volcano, they may block it with disastrous consequences. Besides granite and basalt, there are many other kinds of igneous rocks formed from molten magma. Geologists often classify them according to their silica content into acid (silica-rich), basic and ultra-basic rocks.

Above left: Parícutin, a Mexican volcano, began life on 20 February 1943. Eruptions went on until 1952.
Above right: Lava pours from a fissure in Mount Etna, Sicily, during the 1971 eruption.
Right: The map shows the area covered by ash and the area in which the great 1883 Krakatoa explosion was heard.
Below: Casts of bodies were recovered during excavations at Pompeii.

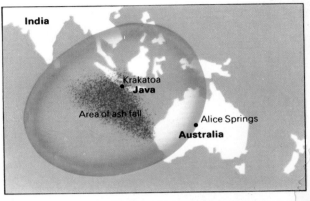

India

Krakatoa
Java

Area of ash fall

Alice Springs
Australia

in 79 AD. The town of Pompeii was buried by volcanic ash and pumice and Herculaneum was overwhelmed by a LAHAR. About half of the dome was destroyed in the eruption, in which no lava flowed. Recent eruptions, including the last in 1944, have been accompanied by lava flows.
Volcanic ash consists of small fragments of magma which are thrown into the air during volcanic explosions. Extremely fine ash is called volcanic dust.
Volcanoes are named after Vulcan, the Roman god of

fire and metal working. They are vents at the Earth's surface through which hot magma, gases and fluids escape. They are broadly

Vesuvius overlooks the Bay of Naples, Italy

divided into explosive, intermediate and quiet types, based on the nature of the eruptions associated with them.

Wegener, Alfred (1880–1930) was the German meteorologist who has been credited as the founder of the theory of CONTINENTAL DRIFT. He advanced the theory in 1912, but his ideas were rejected by most scientists. Born in Berlin, he studied at the universities of Heidelberg, Innsbruck and Berlin. He is supposed to have first thought of his theory while watching a glacier move and float in the sea, during one of his 3 scientific expeditions to Greenland. He had no formal training in geology.

Alfred Wegener

Many mysteries remain to be unravelled about the dark world beneath the oceans and the powerful forces in the atmosphere which cause our changing weather. Further study of both of these zones may be of enormous benefit to mankind.

The Oceans and Weather

Water makes the Earth a rather special planet in the Solar System. The oceans claim over 97 per cent of this great water supply with most of the rest frozen solid in the ice sheets and glaciers. The oceans cover seven out of every ten square kilometres of the planet. They are divided into three main oceans. The vast PACIFIC (covering 30 per cent of the globe) is twice as large as the ATLANTIC (the second largest), which is in turn slightly larger than the INDIAN OCEAN. Two lesser oceans are sometimes also classified – the Antarctic (connected to the above three) and the Arctic (an extension of the Atlantic).

A strange and dark world

It is a strange and still largely unknown world beneath the waves. The average depth of the ocean is 3,800 metres, although one of the deepest parts, the Marianas Trench off the coast of the Philippines, goes down to 11,033 metres – more than 2,000 metres deeper than the height of Mount Everest. The water pressure at the bottom of the trench is more than 300 times greater than at the surface. Therefore, BATHYSCAPHES, underwater scientific vessels, were developed to explore the ocean deeps.

From the coastline of a continent, the very gently sloping continental shelf drops about 1 metre in every 500, but then this increases to 1 in 40 at the edge of the shelf where the continental slope begins, leading down to the ocean's ABYSSAL DEPTHS. However, sediment, thousands of metres thick in places, nearly always blocks the underwater traveller's view of the bottom rocks. Yet the geography of the ocean floor is as varied a landscape as any part of the continents – flat plains, canyons deeper than the Grand Canyon, and "rivers" of moving sediment. The 600,000 kilometres of mid-oceanic ridges are as large and as impressive as the Alps, the Rockies, or even the Himalayas.

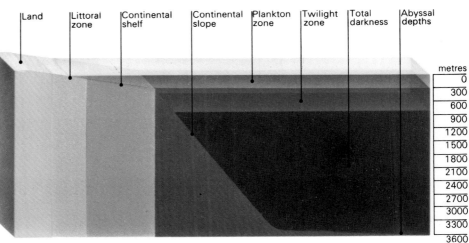

Land | Littoral zone | Continental shelf | Continental slope | Plankton zone | Twilight zone | Total darkness | Abyssal depths

metres: 0, 300, 600, 900, 1200, 1500, 1800, 2100, 2400, 2700, 3000, 3300, 3600

Above: The continental shelf dips gently from the land to the edge of the continental slope, which plunges down to the abyss. Most living organisms are found in the top sunlit layer of the ocean.

Below: The Severn Bore is a surge or high tidal wave of water which rushes up the River Severn in England. It occurs when the high tidal wave slows down as it reaches shallow water and meets the river current. The water piles up and the wall of water may, at spring tide, be a metre or so high.

Reference

Geology and oceans

Life first began in the sea and the oceans have always had an influence on geology and man. Indeed, the Earth's geological history has been dominated by the constantly changing configuration of ocean and continent. The continents were like islands wandering in a great sea.

First of all, the sea influences the weather. The process whereby water, in some form or other, circulates from the oceans to land areas and back again is called the water cycle. It all starts with water evaporating from the sea. Clouds carry this water vapour over the land. This then falls as rain or snow to return eventually to its origin – the sea – via rivers or as groundwater.

The oceans were the starting point for most of the sedimentary rock we find on the continents today. The ocean's sediments – the loose material that eventually becomes sedimentary rock – come from a variety of places. Debris, such as sand and silt, from the continents is washed into the sea by rivers. Volcanoes erupt under the sea. The sea is constantly pounding away at the continents and eroding the coasts. In addition, the rivers carry 8,000 million tonnes of salts into the sea each day! In the sea itself, many creatures have calcareous or silica shells and skeletons. When they die, their remains sink and accumulate on the ocean bottom. Organisms, such as

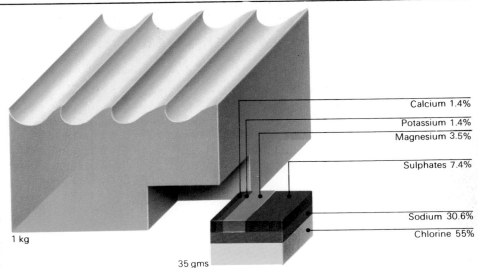

Calcium 1.4%
Potassium 1.4%
Magnesium 3.5%
Sulphates 7.4%
Sodium 30.6%
Chlorine 55%

1 kg

35 gms

Above: The diagram shows that one kilogram of seawater contains 35 grams of dissolved substances, the chief one being common salt, or sodium chloride. But nearly every element is also present in seawater and salt, bromine and magnesium are commercially extracted from it.
Below: Heron Island is part of the Great Barrier Reef, a coral formation off the coast of Australia.

radiolarians, diatoms and coccoliths, accumulate to form oozes which cover more than half the ocean floor.

Limestone and chalk rocks are a collection of millions of skeletons – a stick of schoolroom chalk would probably contain a quarter of a million shells. Larger creatures, for example algae, molluscs, sea urchins and corals, also form limestone. In the future, the modern coral islands in the Pacific and Indian Oceans will themselves eventually form a type of limestone.

However, such processes take literally ages. Although the carpet of sediment on the ocean floor is sometimes 750-1,000 metres thick, the rate of accumulation is very slow. In the Pacific and Indian Oceans the rate is between a millimetre and a centimetre per 1,000 years, while in the Atlantic it is one to ten centimetres every 1,000 years.

Oceans and man

Surface currents in the oceans have been recorded since ancient times and were used by early navigators. The TIDAL rhythms caused by the powerful pull of the Moon's gravity still affect modern shipping. Water CURRENTS affect the climate of countries in their path. The warmth of the Gulf Stream allows palm trees to grow in the extreme south-west of Britain. When warm surface water is blown away from cold water areas, colder water rises to the surface to take its place. Off the coast of Peru, this brings up plankton (minute marine organisms) which are

B **Barometer.** An instrument that measures AIR PRESSURE, either by a column of mercury or by a vacuum chamber (aneroid barometer). An altimeter indicates the height of an aircraft by measuring the air pressure, which decreases with height.

White clouds in 'mackerel' sky

Bathyscaphe is a mobile underwater vehicle developed in the late 1940s. It allows scientists to descend to around 10,000 metres and explore the ABYSSAL DEPTHS in comparative safety.
Beaufort scale measures wind speed. It ranges from Force 1 (causes smoke to drift) to Force 12 (over 120 km/h, or hurricane force).

C **Climate** describes the typical weather for an area. It is based on average statistics, such as for temperature and rainfall, taken over a long period of time. A country's climate depends upon its latitude (polar or EQUATORIAL), proximity to the sea (MARITIME or CONTINENTAL), and type of terrain.
Cloud seeding means dropping dry ice or silver iodide into clouds to make rain.
Clouds are masses of tiny condensed water droplets, or ice crystals. Warm air can hold a great deal of invisible moisture compared to cold air. When warm air cools, moisture is condensed into visible droplets. Clouds come in a great many shapes and sizes and at a variety of heights: low clouds (below 2,500 metres) are Stratus, Stratocumulus, Nimbostratus, and Cumulus and Cumulonimbus (storm clouds). Medium clouds (between 2,500–6,100 metres) are Altostratus and Altocumulus. High clouds (above 6,100 metres) are Cirrocumulus, Cirrostratus and Stratus.
Condensation is the change in water, on cooling, from invisible water vapour into liquid droplets.
Continental climate. This type occurs in the interiors of the large continents. It is

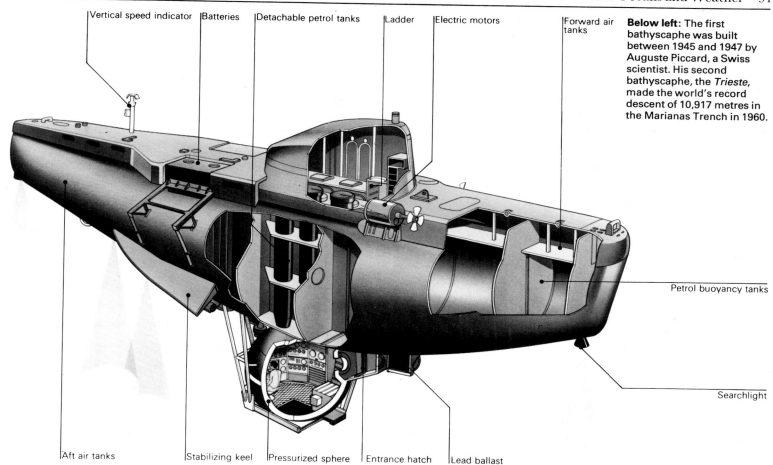

Vertical speed indicator | Batteries | Detachable petrol tanks | Ladder | Electric motors | Forward air tanks

Below left: The first bathyscaphe was built between 1945 and 1947 by Auguste Piccard, a Swiss scientist. His second bathyscaphe, the *Trieste*, made the world's record descent of 10,917 metres in the Marianas Trench in 1960.

Petrol buoyancy tanks

Searchlight

Aft air tanks | Stabilizing keel | Pressurized sphere | Entrance hatch | Lead ballast

Left: Storm waves break against the volcanic cliffs of Lanzarote in the Canary Islands. Most waves are caused by winds that blow across the open sea.

Below: In the 1960s, 2 volcanic islands appeared off Iceland, Surtsey, and Little Surtsey (shown here). Little Surtsey was destroyed by constant wave action.

A tropical rain forest

noted for its extremes of hot and cold temperatures, both daily and seasonally. Other features include a low rainfall and low HUMIDITY.

Continental shelf is the gently sloping part of the sea floor at the edge of continents. It is never more than 200 metres deep.

Coriolis force is the bending effect of the Earth's rotation on its wind systems. In the Northern Hemisphere, the 'twist' of the winds around a low pressure zone (a depression) forms an anticlockwise movement, changing to clockwise around a high pressure zone (anticyclone). The exact reverse occurs in the Southern Hemisphere.

Currents in the oceans are the regular movement of warm or cool surface water. They often moderate the climates of coastal areas, and are responsible for the gene-

ral circulation of seawater. Currents caused by prevailing winds are known as drift currents.

D Depression, or cyclone, is an area of low pressure.

E Echo sounder, or sonar, is a device to measure the depth and shape of the ocean bottom. Sound waves are bounced off the bottom and the time taken for their return measures the distance travelled. They are also used to locate submarines and fish shoals.

Equatorial climate. This is characterized by its heavy rainfall and high temperatures all year round. Rain forest is the typical vegetation. It is limited to areas 5°N or S of the equator.

Equinox is the time of year, around March 21 and September 22, when the Sun is overhead at the equator and night and day are of equal length.

F Fog is a dense low-lying cloud, i.e. a suspension of small water droplets in the air. Polluted fog may become smog.

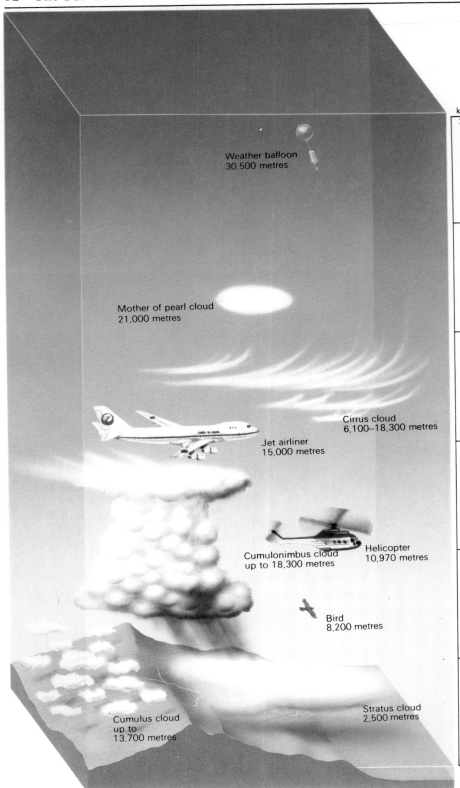

Weather balloon
30,500 metres

Mother of pearl cloud
21,000 metres

Cirrus cloud
6,100–18,300 metres

Jet airliner
15,000 metres

Cumulonimbus cloud
up to 18,300 metres

Helicopter
10,970 metres

Bird
8,200 metres

Stratus cloud
2,500 metres

Cumulus cloud
up to
13,700 metres

Left: The diagram shows the atmosphere up to 30,500 metres above the ground. The air becomes increasingly thinner as one ascends. In fact, 75% of the mass of the atmosphere and 90% of its moisture are in the lowest zone, the troposphere. Most of the weather conditions felt on Earth occur in the troposphere.

an attractive lure for great shoals of fish. It also makes the climate colder and brings fog to coasts where the cool sea air meets the warm air from the land. Seawater is also a vast resource of mineral salts in solution, while other minerals, such as diamonds, coal and petroleum, are mined in offshore areas.

The weather

Dr. Samuel Johnson, the English sage, once rightly said, "When two Englishmen meet, their first talk is of the weather." The weather influences the food we eat, the way we dress and the houses we live in. Weather itself depends on a combination of TEMPERATURE, AIR PRESSURE, air speed, moisture content, CLOUDS, and PRECIPITATION. How does it all work?

The Earth rotates around the Sun. Our Sun is not unusual since there are about 100,000 million of them in our galaxy alone. It is special to us, however, since it gives out heat, light, and supports life. It is a vast ball of exploding gases (mainly hydrogen that converts to helium through a nuclear reaction) which reaches a temperature of 6,000°C at the surface.

The heat and light have to travel almost 150 million kilometres to the Earth, passing Mercury and Venus on the way. The journey is simple until it reaches the Earth's ATMOSPHERE, a thin shell of gases surrounding the Earth. This acts as a very effective protective shield which reflects and absorbs harmful radiation and objects like meteorites. The problems that rockets have on re-entering the Earth's atmosphere indicate what an effective barrier it forms.

An extremely important layer in the atmosphere lies between 25 and 50 kilometres above the ground. This is rich in a gas called OZONE. Like giant sunglasses, this layer filters out the deadly ultra-violet light from the Sun. Below this the atmosphere becomes much more mixed. The

Altostratus cloud

Föhn winds are warm, dry winds created when a prevailing wind has lost most of its moisture by being forced up a mountain range. As it blows down the other side, it becomes warmer. The best known example is the Chinook ('snow eater') of the Rocky Mountains in North America.
Fronts are zones separating warm and cold AIR MASSES.
Frost occurs whenever the air temperature reaches freezing point (0°C). The water droplets condense as ice on surfaces such as grass and windows.

Hailstones are small chunks of ice that form within a cloud by the gradual accumulation of water around a particle of dust. Hail falls when these particles become too heavy to be carried by rising air currents.
Humidity is the concentration of water vapour in the ATMOSPHERE. The maximum is about 4%. The relative humidity measures how close damp air is to saturation point (100%).
Hurricanes are destructive whirlpools of air formed over the tropical areas of the world. They are rotating air

masses, the calm centre of which is called the 'eye'. Other names given them include typhoon, Willy Willy, Baguio and tropical cyclone.

Icebergs are huge blocks of floating ice; only 10 per cent of their mass is visible above the water. They are usually formed when the

Hurricane damage in Mississippi, USA

Below: The diagram shows ways in which rain is formed. In each case, the air rises and cools. Hence, its ability to hold water vapour decreases. Cyclonic rain (1), occurs in depressions when warm air rises over a cold air mass. Convectional rain (2), occurs when the Sun heats the land and warm air rises in convection currents. Orographic rain (3) occurs when air masses rise over mountains.

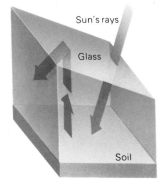

Sun's rays

Glass

Soil

Above: Delicate clouds fill the sky above the Low Tatra mountains in Czechoslovakia.
Left: The Earth's atmosphere acts much like the glass in a greenhouse. Heat energy penetrates through the atmosphere, but when it is reflected from the ground, it cannot escape, especially when there is cloud cover. The air, therefore, retains heat because of the so-called 'greenhouse effect'.

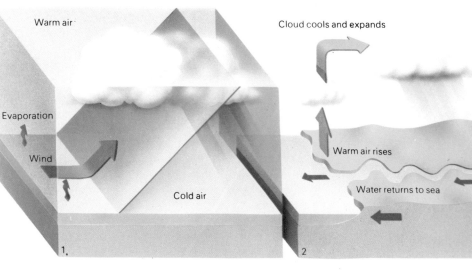

Warm air

Evaporation

Wind

Cold air

1.

Cloud cools and expands

Warm air rises

Water returns to sea

2

Wind

ndwater seepage

good reflectors – the ...eal weather exists ...rs surface. Clouds are very ... and ice on the ground. In the end, only one in every 2,000 million parts of the Sun's energy reaches the Earth.

Recent space voyages have shown what it is like on planets without an atmosphere; hot enough during the day to fry an egg on the rocks and cold enough at night to keep frozen food solid. It has been calculated that if the Earth's atmosphere would disappeared, temperatures at the equator would reach 80°C during the day and fall to –140°C at night.

Windy weather

Air has weight and therefore presses down on the ground. This AIR PRESSURE is greatest at the ground. As we go up, or as air rises moving upwards, the pressure decreases as the becomes thinner. Air pressure also changes ...zontally. When air is heated it expands, ...mes less dense (that is, lighter), and rises. ...is how the great balloons and Zeppelins ...ps) got off the ground – the air in the ...was heated and 'became lighter than air' ...balloons floated upwards. When air is ...mes dense and sinks. Low pressure air systems are

g... M... air... hori... beco... This... (airsh... balloon... and so the ... cooled, it beco...

When air rises,...

---

ends of glaciers calve, or break off, into the sea. They are a danger to shipping, especially in the North ATLAN-TIC. The passenger liner

An iceberg

Titanic was holed by one in 1912, and sank with a loss of 1,513 lives.
Indian Ocean is the small-est of the world's 3 great oceans. Its area is 73 million km². North of the equator, the area is influenced by the Monsoon (*see page 94*).
Isobar is like a contour line, but on a weather map. It connects places having the same AIR PRESSURE. Isotherms connect places having equal temperatures.

J **Jet streams** are belts of fast moving wind, travel-ling at speeds of up to 500

km/h. They occur at high altitudes and mark the edges of fronts between AIR MASSES.

K **Köppen,**Wladimir(1846–1940), a Russia... meteorologist, compiled th... most widely used clima... classification. It is based ...ts or annual and monthly m...ts or temperatures and prec...ely... tion.

L **Latitude** is the di... measured in de... or S of the equator... Tropic of Cancer lie... and the Tropic of ... at 23½°S. Longi...

...and between ...e ground. A ...undreds of ...aged be... Lightning ...current ...uickest ...thun-

...as ...West ...Trad... bring... and ... sides ... areas... inclu... and ... **Mic...** form... gard... area... clim...

...ning over Phoenix, USA

Above: The tops of thunder-clouds acquire a positive electrical charge, while the bottoms are negatively-charged. The discharge is visible as a stroke of lightning. Lightning may travel to the ground, because the bottoms of clouds and the ground have opposite charges. Along the lightning's path, the air is heated. Warm air molec... collide with cold ones... create the sound of...
Right: Rainbows a... the Sun's rays a... and refracted... raindrops i... primary... orange... indig... so...

...lected by being sucked into low pressure zones. One fine example of this is the Monsoon, a wind that brings rain to the otherwise dry areas of Asia. In summer, the Asian landmass heats up more quickly than the Indian Ocean. This creates a low pressure cell over Asia. Winds flow across the Indian Ocean towards Asia, picking up water vapour which falls as the heavy Monsoon rains over the Asian continent. The winter wind system is a complete reversal of this.

The major world wind systems are (from the equator and moving polewards): the low pressure Doldrums (very little wind at all), the North-East and South-East Trades (the direction in the name of a wind is always where it is moving *from*), the high pressure Horse latitudes (so-called because sailors of becalmed ships threw horses overboard to save valuable drinking water), the Westerlies, and the Polar Easterlies. These winds carry with them another very important part of the weather – precipitation.

Rain, hail, snow and fog

Water is unique. It is the only substance that can exist as a liquid (water), solid (ice), and gas (water vapour) at normal temperatures and pressures. Its familiar forms are rain, hail and snow, and fog. The amount of moisture in the air is called the HUMIDITY. When there is too much water in the air it often condenses into a mass of

---

Maritime climate. This type is influenced [by th]e sea, thus having cool [summ]ers and mild winters. **[Med]iterranean climate.** [...] is dominated by the [West]erlies in winter and the [Trad]es in summer. These [give] warm, dry summers [and m]ild, damp winters. Be[...] southern Europe, other [areas] that enjoy this climate [inclu]de parts of California [and S]outh Africa. **[Micr]oclimate** is a localized [area] of CLIMATE. A sheltered [gard]en, a lake or a built-up [area] can all generate local [clim]atic conditions that differ from the prevailing regional climate.

O **Ooze,** see ABYSSAL DEPTHS.
Ozone (O_3) is a gas present in the upper atmosphere that protects the Earth from the Sun's ultra-violet radiation.

P **Pacific Ocean.** The largest ocean in the world, it was so-named by Ferdinand Magellan, the Portuguese navigator, because he had a calm crossing in 1521. It has an area of 165 million km² and is some 16,000 km N-S. It also contains the deep ocean trenches: the Marianas Trench goes down to 11,033 metres.
Permafrost is permanently frozen ground. It occurs extensively in polar and tundra regions where the average annual temperature is below 0°C and less than 25 cm of rain falls.
Precipitation is moisture from the air which falls as RAIN, HAIL or SNOW and forms part of the WATER CYCLE.

R **Rain** is condensed moisture falling as drops from clouds.

Rain gauge is one of the simplest and most ancient of weather instruments. It consists of a funnel and a glass jar which has a scale in millimetres. The rain falling into the jar is measured twice daily.
Rainbows are the coloured arches in the sky formed by sunlight scattered into its basic colours by raindrops. The colours are always in the order of red, orange, yellow, green, blue, indigo, and violet. Sometimes secondary bows occur with these colours in reverse order.

good reflectors, as are water surfaces and snow and ice on the ground. In the end, only one in every 2,000 million parts of the Sun's energy reaches the Earth.

Recent space voyages have shown what it is like on planets without an atmosphere; hot enough during the day to fry an egg on the rocks and cold enough at night to keep frozen food solid. It has been calculated that if the Earth's atmosphere disappeared, temperatures at the equator would reach 80°C during the day and fall to −140°C at night.

Windy weather

Air has weight and therefore presses down on the ground. This AIR PRESSURE is greatest at sea level. Moving upwards, the pressure decreases as the air becomes thinner. Air pressure also changes horizontally. When air is heated it expands, becomes less dense (that is, lighter), and rises. This is how the great balloons and Zeppelins (airships) got off the ground – the air in the balloon was heated and 'became lighter than air' and so the balloons floated upwards. When air is cooled, it becomes dense and sinks.

When air rises, low pressure air systems are

Below: The diagram shows ways in which rain is formed. In each case, the air rises and cools. Hence, its ability to hold water vapour decreases. Cyclonic rain (1), occurs in depressions when warm air rises over a cold air mass. Convectional rain (2), occurs when the Sun heats the land and warm air rises in convection currents. Orographic rain (3) occurs when air masses rise over mountains.

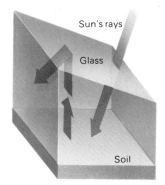

Sun's rays
Glass
Soil

Above: Delicate clouds fill the sky above the Low Tatra mountains in Czechoslovakia.
Left: The Earth's atmosphere acts much like the glass in a greenhouse. Heat energy penetrates through the atmosphere, but when it is reflected from the ground, it cannot escape, especially when there is cloud cover. The air, therefore, retains heat because of the so-called 'greenhouse effect'.

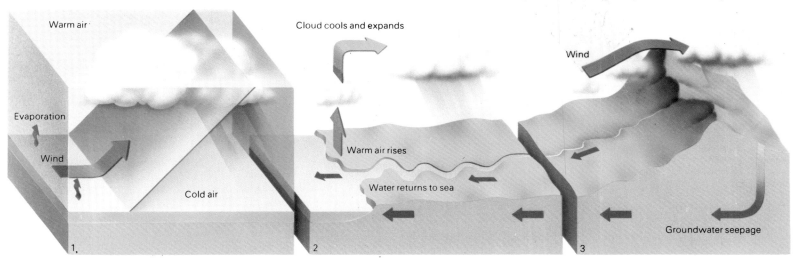

Warm air
Cloud cools and expands
Wind
Evaporation
Wind
Cold air
Warm air rises
Water returns to sea
Groundwater seepage
1.   2.   3.

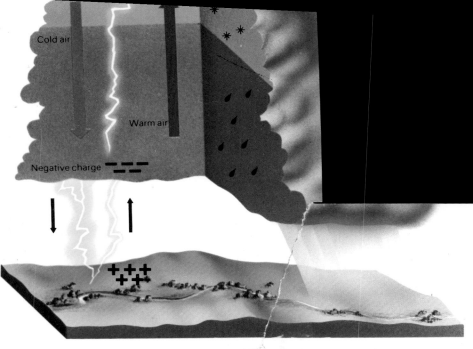

Above: The tops of thunder-clouds acquire a positive electrical charge, while the bottoms are negatively-charged. The discharge is visible as a stroke of lightning. Lightning may travel to the ground, because the bottoms of clouds and the ground have opposite charges. Along the lightning's path, the air is heated. Warm air molecules collide with cold ones to create the sound of thunder.
Right: Rainbows occur when the Sun's rays are reflected and refracted (bent) by raindrops in the air. The primary bow is coloured red, orange, yellow, green, blue, indigo and violet, although some colours may be blurred. In secondary bows, seen in the top right of the picture, the colours are reversed.

flected by being sucked into low pressure zones. One fine example of this is the Monsoon, a wind that brings rain to the otherwise dry areas of Asia. In summer, the Asian landmass heats up more quickly than the Indian Ocean. This creates a low pressure cell over Asia. Winds flow across the Indian Ocean towards Asia, picking up water vapour which falls as the heavy Monsoon rains over the Asian continent. The winter wind system is a complete reversal of this.

The major world wind systems are (from the equator and moving polewards): the low pressure Doldrums (very little wind at all), the North-East and South-East Trades (the direction in the name of a wind is always where it is moving *from*), the high pressure Horse latitudes (so-called because sailors of becalmed ships threw horses overboard to save valuable drinking water), the Westerlies, and the Polar Easterlies. These winds carry with them another very important part of the weather – precipitation.

Rain, hail, snow and fog

Water is unique. It is the only substance that can exist as a liquid (water), solid (ice), and gas (water vapour) at normal temperatures and pressures. Its familiar forms are rain, hail and snow, and fog. The amount of moisture in the air is called the HUMIDITY. When there is too much water in the air it often condenses into a mass of

---

Lightning over Phoenix, USA

M **Maritime climate.** This type is influenced by the sea, thus having cool summers and mild winters.
Mediterranean climate. This is dominated by the Westerlies in winter and the Trades in summer. These bring warm, dry summers and mild, damp winters. Besides southern Europe, other areas that enjoy this climate include parts of California and South Africa.
Microclimate is a localized form of CLIMATE. A sheltered garden, a lake or a built-up area can all generate local climatic conditions that differ from the prevailing regional climate.

O **Ooze,** see ABYSSAL DEPTHS.
Ozone (O_3) is a gas present in the upper atmosphere that protects the Earth from the Sun's ultra-violet radiation.

P **Pacific Ocean.** The largest ocean in the world, it was so-named by Ferdinand Magellan, the Portuguese navigator, because he had a calm crossing in 1521. It has an area of 165 million km² and is some 16,000 km N-S. It also contains the deep ocean trenches: the Marianas Trench goes down to 11,033 metres.
Permafrost is permanently frozen ground. It occurs extensively in polar and tundra regions where the average annual temperature is below 0°C and less than 25 cm of rain falls.
Precipitation is moisture from the air which falls as RAIN, HAIL or SNOW and forms part of the WATER CYCLE.

R **Rain** is condensed moisture falling as drops from clouds.

Rain gauge is one of the simplest and most ancient of weather instruments. It consists of a funnel and a glass jar which has a scale in millimetres. The rain falling into the jar is measured twice daily.
Rainbows are the coloured arches in the sky formed by sunlight scattered into its basic colours by raindrops. The colours are always in the order of red, orange, yellow, green, blue, indigo, and violet. Sometimes secondary bows occur with these colours in reverse order.

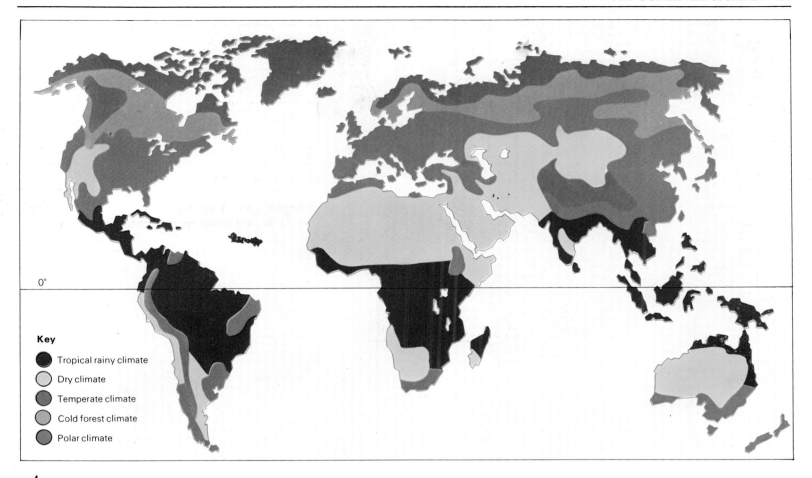

Key

- Tropical rainy climate
- Dry climate
- Temperate climate
- Cold forest climate
- Polar climate

0°

tiny droplets. Near the ground this causes fog or mist, but higher up it forms clouds. These come in all shapes and sizes. As long as the tiny droplets or ice crystals stay separate then the cloud merely drifts along, perhaps joining with others to make vast layers of moisture in the air.

Often, however, larger droplets gather smaller ones and begin to grow. The same thing happens if a piece of dust or ice falls through the cloud – it gathers droplets around it and grows. Eventually it becomes so large and heavy, perhaps a million times larger than the original droplet, that it falls out of the cloud as rain, snow, or hail. This is precipitation. Precipitation is also formed when a cloud is forced to rise, perhaps to pass over a range of mountains. As it does so, it cools and cannot hold as much moisture. Cool air can hold less water vapour than warm air. It therefore unloads its excess moisture by condensation into rain, snow or hail.

Above: The map shows a simple classification of world climates devised by the Russian meteorologist Wladimir Köppen.

Below: Near the poles, the Sun's rays pass through more atmosphere and spread over a larger area than at the equator. Hence, equatorial zones are warmer than polar areas.

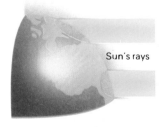

Sun's rays

Fronts also cause precipitation. These are the zones along which warm (less dense) air masses meet cold (dense) air masses. The warmer air rises over the cold air. This is called a warm front. The upward movement causes cooling and thus precipitation.

Violent weather

The most common form of violent weather is a THUNDER and LIGHTNING storm – over 45,000 occur daily all over the world. Less common, but much more destructive, are the tropical cyclones originating in the tropical areas (5°–20° either side of the equator). Each area has its own name for them – hurricane in North America, typhoon in eastern Asia, cyclone in India, Willy Willy in Australia, and Baguio in the China Sea area.

The power of these storms is enormous; in just one day a hurricane produces 200 times more energy than is produced in the USA, and drops

Moonlight on the sea

S **Seasons** of the year are spring, summer, autumn (fall in North America), and winter. In December, it is summer in Australia but winter in Europe. The seasons are caused by the Earth's axis being tilted and the overhead Sun moving north or south of the equator.
Snow is frozen water vapour, consisting of tiny ice crystals, that falls to the ground in flakes.
Solstices are the times of year when the overhead Sun reaches its northernmost and southernmost points.

The summer solstice in the Northern Hemisphere occurs on June 21; the Sun is then overhead at the Tropic of Cancer. At the winter solstice, December 21, the Sun is overhead at the Tropic of Capricorn. *See* LATITUDE.

T **Temperature** is measured by a thermometer. The degree of heat is expressed by either the Centigrade scale (°C) or the Fahrenheit scale (°F).
Thunder is a deep rumbling sound in the air. It is formed when LIGHTNING heats up the

surrounding air to about 15,000°C (compared to the Sun's surface temperature of 6,000°C). The position of a thunderstorm can be calcu-

lated by counting the number of seconds between the flash and the clap of thunder. Roughly 3 seconds equals 1 km.

Tornado over Minnesota, USA

Key
- Rain
- Snow
- Sleet
- ⑤ Temperature (°C)
- ② Wind speed (km/h)
- Cold front
- Warm front

High

1008 1004 1000 996 992 988 984 **Low**

1020

1016

1012

1016

1012

1016

1020

High

High

around 10 to 20,000 million tonnes of water. It is thought that they are created by warm surface air rising very rapidly in a narrow ring-shaped pillar of cloud. Air from hundreds of kilometres around rushes in a spiral motion towards the centre, accelerating all the time.

Anybody unfortunate enough to be caught in a hurricane would go through a series of strange events as the storm passed him by. As the air pressure falls, the normal 15–26 kilometres an hour (km/h) winds would increase to gale force at around 40 km/h. Nearer the storm's centre the winds would increase enormously to reach over 300 km/h and the rain would become torrential. At the centre, called the eye, things become calm, winds return to 15–20 km/h and the sun often shines through the surrounding wall of cloud. This is only the lull before the storm. Suddenly the second half of the storm will begin, just as fierce as the first, but everything is in the reverse order and blowing in the opposite direction!

Above: Weather maps are drawn from data received from weather stations. The above map shows wintry conditions over western Europe, with a low pressure system, or depression, over Scandinavia and a high pressure system, or anticyclone, over south-western France. A cold front, associated with rain, is over Britain. The map also shows temperatures and places affected by rain, snow and sleet. When the data arrives at a forecast centre, experts prepare a synoptic chart, a weather map which gives a synopsis or summary of the weather at a particular time. By comparing this chart with earlier charts, forecasters prepare a prognostic chart which shows how they think the weather will develop in the near future.

Distribution and forecasting

Although the weather of the world follows a certain general pattern, there are too many factors influencing it to set any definite rules. Land and sea distribution, mountains, ocean currents, the Earth's rotation, and large industrial towns pumping out heat, all make the weather picture very complicated. There are a number of climatic types – MEDITERRANEAN, polar, CONTINENTAL, MARITIME, and EQUATORIAL – but they certainly do not exist in straight-sided belts parallel to the equator.

All these complicating factors make life very difficult for weather forecasters. The need to forecast is obvious; from transport schedules, through farmers' growing seasons, to sporting events, it is important to know the future weather. Despite weather ships, balloons, aircraft, and now satellites collecting data and feeding it into computers, the weather often remains as unpredictable as ever.

Tides are the regular rise and fall of the sea. On average, they occur every 12 hours. They are caused by the gravitational attraction of the Moon and, to a lesser extent, the Sun. When the Earth, Sun and Moon are in a straight line, the attraction is the strongest. This causes high Spring tides. When the Sun and Moon are at right angles to the Earth, the attraction is less. These low tides are called Neap tides.
Tornadoes (from the Spanish *tornar*, to turn) are intense and violent spirals of air about 0.4 km across. The

wind speeds vary between 200–400 km/h. They are often very destructive and occur mainly east of the Rockies in the USA, the southern USSR and Australia. About 150 a year occur in the USA.

W **Water** is the basis of all life. It is unique in that it exists in 3 natural states – liquid (water), solid (ice), and gas (water vapour).
Water cycle is the process whereby water, in one form or another, circulates from the oceans to the land and back again.

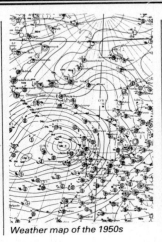
Weather map of the 1950s

Waterspout is a funnel-shaped mass of cloud and water, similar to a TORNADO but occurring at sea.
Weather is the combination of sunshine, rain, and wind. The science of weather is called meteorology. For the first time the use of satellite pictures gives a complete view of the world's weather systems from above – which means that a storm can be tracked around the world.
Weatherlore is the way in which people formerly forecasted the weather. Usually, this was by way of a saying or a particularly well-

known rhyme, eg:
Red sky at night,
       sailors' delight.
Red sky in morning,
       sailors take warning.
Wind is the movement of air caused by air rising or sinking. People depended on the wind in the age of sail to aid their travel and trade, so winds were one of the earliest weather systems to be studied. The world's wind systems can provide steady winds or gales – the Roaring Forties are strong south-westerly winds. Wind on water causes waves.

Although difficult to observe in one person's lifetime, land areas are constantly planed down to almost level surfaces. The chief agents of denudation are wind and weather, running water, large, moving bodies of ice and sea waves.

Natural Landscapes

Above: The bare rock landscape in the Bryce Canyon National Park of Utah, in the USA, reveals the power of natural erosion.

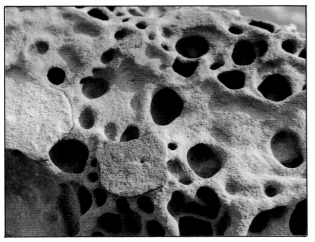

Left: The pitted surface of this oolitic limestone rock is an erosional feature caused by solution.

The inscriptions on gravestones more than two or three hundred years old are usually illegible, because they have been worn away by rain, wind and frost. The Earth itself is being worn down in much the same way. It is under constant attack. Geologists estimate that, on average, about 3.5 centimetres of land are worn away from land surfaces every 1,000 years.

As soon as any rock is exposed on the Earth's surface, it is assaulted by the processes of WEATHERING and the agents of EROSION. Weathering, whether mechanical or chemical, takes place *in situ*, that is, it starts and finishes in one place. However, the agents of erosion – rivers, glaciers, the sea, winds – transport rock fragments from one place to another. In so doing, the rock debris is used to erode even more material. For example, waves hurl loose rocks against cliffs and undercut them.

The worldwide cycle of weathering, transportation and erosion (also called DENUDATION) is both destructive and constructive. It constantly changes the face of the Earth by breaking rocks down and redistributing the debris to form new rocks and landscapes.

Landscape and climate affect this cycle — weathering is most rapid in mountainous areas. As soon as a rock is weakened and dislodged from the rest, it falls by gravity. A new surface is then exposed to the elements. Climate dictates water supply, temperatures and winds — the forces at work in arid regions differ from those in moist regions.

The speed at which weathering and erosion occur depends largely on the type of rock under attack. Massive igneous rocks can usually resist weathering for long periods. However, even granite decomposes into soft kaolin (CHINA CLAY) over millions of years. On the other hand, a relatively soft limestone is dissolved away fairly rapidly by rainwater.

Reference

A **Abrasion** is the process by which running water, bodies of ice and winds use eroded fragments of rocks to wear away the land.

Aeolian deposits are the accumulation of wind-blown sediments, such as sand DUNES and LOESS.

Aletsch Glacier, in southern Switzerland, is the largest glacier in the Alps. Made up of the Great, Upper and Middle Aletsch Glacier, the largest section is 2,643 metres above sea level. The glacier is on the 4,198-metre high mountain called the Aletschhorn.

Alluvial deposits, mainly sands and gravels, are transported and deposited by rivers.

Alluvial fan is a wedge-shaped, roughly triangular deposit created when a river slows down because of a change of gradient. Several alluvial fans united together form a piedmont plain.

Amazon River, in South America, is the world's second longest after the Nile. It is 6,570 km long and it has the world's largest volume of water. The Amazon and its tributaries drain a vast basin, covering 7,045,000 km², which is nearly as large as Australia. The source, a stream named Huarco which rises near the summit of Cerro Huagra, a mountain in Peru, was disco-

Bernina glacier, Switzerland

vered only in 1953. Ships can travel up the Amazon for about 3,800 km.

Antarctic ice sheet covers most of the south polar continent. It covers about 13 million km² and contains about 25 million km³ of ice. The maximum thickness of the ice is about 4 km.

Aquifer is a water-bearing layer of rock, such as sandstone, gravel or chalk.

Arête, see CIRQUE.

Artesian structures are those in which water is held underground in a tilted or saucer-shaped AQUIFER. The sponge-like aquifer is usual-

Mechanical and chemical weathering

One type of mechanical weathering is called exfoliation, in which the outer layers of exposed rocks are peeled away like the skins of an onion. This process occurs in deserts, because of the rapid heating during the day followed by fast cooling at night.

Water is important in mechanical weathering, especially when it fills cracks in rocks or is absorbed by minerals. When water turns into ice, it expands, occupying a larger volume than the original water. The force of this expansion loosens and eventually shatters rocks. This process creates large piles of loose rock called SCREE, which often accumulate at the bottoms of mountain slopes.

The roots of trees and other plants also exert pressure and widen cracks in rocks. Burrowing animals, such as worms, ants and moles, also loosen rocks and bring material to the surface. Even man contributes to mechanical weathering, through road-building, farming, mining and quarrying, and so on.

In chemical weathering, rocks are decomposed (broken down) and destroyed by chemical reactions, although new minerals may be created elsewhere from the same material.

Although pure water dissolves minerals such as salt, rainwater is even more powerful. It absorbs materials such as carbon dioxide (from the air and soil) or acids from rotting plant and animal material and becomes a weak, but effective carbonic acid. As it percolates through the ground, it can attack minerals, dissolve them and carry them away, perhaps to redeposit them elsewhere. This process is particularly marked in limestone (*see pages 104-105*).

Another example of chemical weathering occurs when oxygen in water combines with, or oxidizes, minerals in rocks to form compounds called oxides. For example, when iron in rocks is oxidized, a weak, flaky red rust called iron oxide is formed. Being weak, it is more easily removed by the agents of erosion.

Soils

Soil is a familiar substance, but it is vital, not only to the many organisms that live in it, but also to any animal that eats plant food. A soil starts to form when microscopic organisms called bacteria, and small plants, such as mosses, begin

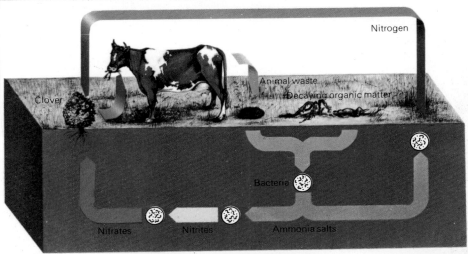

Nitrogen
Clover
Animal waste
Decaying organic matter
Bacteria
Nitrates   Nitrites   Ammonia salts

Above: The diagram shows the nitrogen cycle, whereby nitrogen is constantly redistributed from the soil through plants and animals and then back again through manure or through decomposing carcases. Plants cannot absorb nitrogen directly from the air. They take it from the soil, where it is deposited by rain or taken up by bacteria.
Left: Vines in France are sprayed to protect them against harmful pests and diseases. Farming has become increasingly scientific, so that better crops and larger harvests result. It is also important in protecting the soil's fertility.

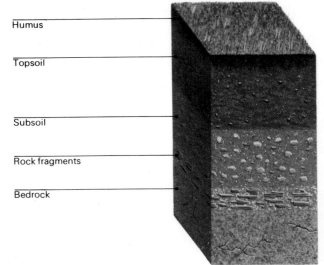

Humus
Topsoil
Subsoil
Rock fragments
Bedrock

Left: Soils contain several distinctive layers, or horizons. The top layer, or A horizon (topsoil), contains fine rock particles and humus from dead plants and animals. The B horizon (subsoil) has little humus and the C horizon (decomposed rock) consists of partly decomposed rock grading down to solid rock.

---

ly sandwiched between water-tight layers, such as clay. At least one end of the aquifer is exposed at the land surface so that rain-water can enter. The buried water is usually under pressure so that when a well is drilled into the aquifer, the water often gushes up like a fountain.
Attrition is the process by which rock fragments are worn into smaller and smaller pieces during their transportation by running water, ice bodies or winds.
Avalanches are falls of snow, ice and rock down

mountain slopes. They often occur when the snow starts to melt in the spring. Others are caused by earthquakes or by loud sounds, such as pistol shots.
Ayers Rock is a large INSELBERG, about 320 km south-west of Alice Springs, in the heart of Australia. It is 2.5 km long and 360 metres high. It rises from the centre of an almost flat plain.

B Bar is a ridge of sand, gravel and pebbles which is usually parallel to the shore. It may enclose a bay.

Base level is an imaginary extension of sea level beneath the land surface. Rivers gradually wear down the land to base level, where there is no longer any slope and so river erosion stops. Land at base level is often called a PENEPLAIN.
Basin is a depression in the Earth's surface caused by erosion, the weight of covering rock or ice, or by the downwarping of rocks.
Bedding planes separate one layer of rock from another. They are lines of weakness, along which rocks can be eroded.

Bedrock is the solid, un-weathered rock lying beneath the soil.
Bergschrund is a CREVASSE near the source of a glacier.

It is created as the ice moves out of a CIRQUE.
Black earth, see CHERNOZEM.
Bog bursts occur when the 'drains' of a bog or swamp

Sand bar, Hawkesbury River, New Zealand

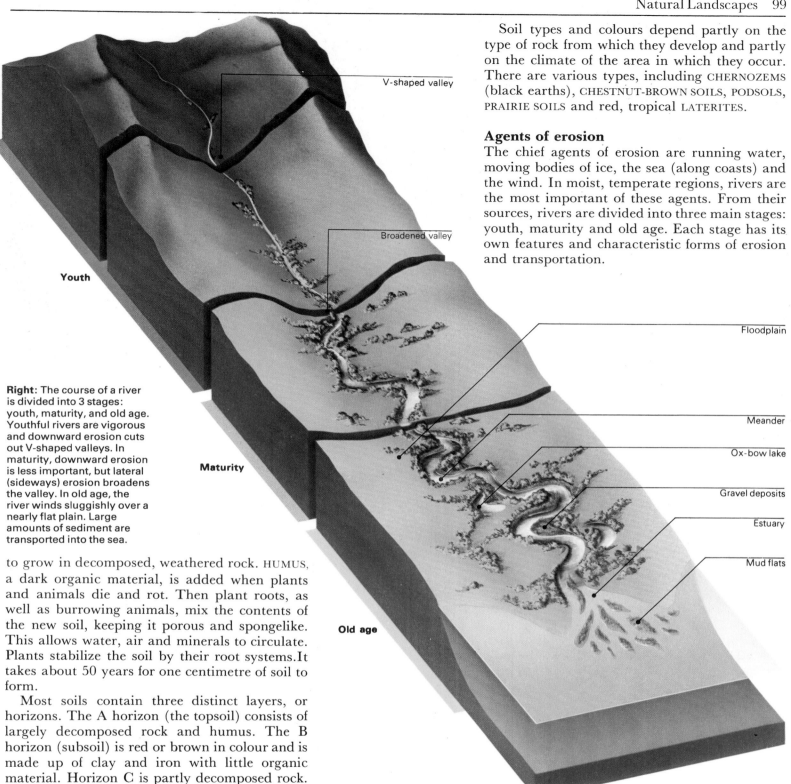

V-shaped valley

Youth

Broadened valley

Maturity

Old age

Floodplain

Meander

Ox-bow lake

Gravel deposits

Estuary

Mud flats

Soil types and colours depend partly on the type of rock from which they develop and partly on the climate of the area in which they occur. There are various types, including CHERNOZEMS (black earths), CHESTNUT-BROWN SOILS, PODSOLS, PRAIRIE SOILS and red, tropical LATERITES.

Agents of erosion

The chief agents of erosion are running water, moving bodies of ice, the sea (along coasts) and the wind. In moist, temperate regions, rivers are the most important of these agents. From their sources, rivers are divided into three main stages: youth, maturity and old age. Each stage has its own features and characteristic forms of erosion and transportation.

Right: The course of a river is divided into 3 stages: youth, maturity, and old age. Youthful rivers are vigorous and downward erosion cuts out V-shaped valleys. In maturity, downward erosion is less important, but lateral (sideways) erosion broadens the valley. In old age, the river winds sluggishly over a nearly flat plain. Large amounts of sediment are transported into the sea.

to grow in decomposed, weathered rock. HUMUS, a dark organic material, is added when plants and animals die and rot. Then plant roots, as well as burrowing animals, mix the contents of the new soil, keeping it porous and spongelike. This allows water, air and minerals to circulate. Plants stabilize the soil by their root systems. It takes about 50 years for one centimetre of soil to form.

Most soils contain three distinct layers, or horizons. The A horizon (the topsoil) consists of largely decomposed rock and humus. The B horizon (subsoil) is red or brown in colour and is made up of clay and iron with little organic material. Horizon C is partly decomposed rock. It grades down to solid, unaltered rock.

---

become blocked, causing it to overflow.
Boulder clay, or till, is a glacial deposit consisting of pebbles and boulders set in

A braided river

thick clay. Many of the rock fragments are polished and marked by being dragged along in the ice.
Braided streams are channels of water weaving in and out of banks of soft sediment. They usually occur when fast-flowing water reaches a flat area containing sandbanks or alluvium deposits, such as a DELTA.
Brown forest soils, or brown earths, are rich in HUMUS formed from decomposed leaves from deciduous trees. The soils are usually dark in colour and fertile.
Butte, see MESA.

A stalagmite

C **Canyon** is a deep, steep-sided valley, resulting from river erosion.
Carlsbad caves, in New Mexico, USA, contain the Big Room, the world's largest underground chamber, 400 metres below the surface. It is 1,220 metres long, 200 metres wide and nearly 100 metres high.
Caverns are underground chambers. The deepest is in the Pyrenees area of France and called the Reseau de la Pierre St Martin. It reaches a depth of 1,332 metres.
Chernozems, or black earths, are extremely rich,

dark soils found in temperate grasslands. The richness comes from the grass which decomposes into HUMUS. It can accumulate because the rainfall in these regions is moderate.
Chestnut-brown soils are similar to CHERNOZEMS, but they occur in drier regions.
China clay, or kaolin, is a pure form of clay formed by the decomposition of feldspar in granite.
Cirque is a semi-circular hollow in which glaciers are formed. Cirques are called corries in Scotland and cwms in Wales. Cirques

Ground water, wells and springs

Rivers start life in several ways. They may flow from a lake or from a melting glacier, or they may emerge from a spring. The latter often occur at the base of hills, where a layer of water-bearing rock reaches the surface.

When rain falls on the land, some of it is evaporated, some flows directly over the surface as run-off, and some seeps through the soil into the rocks below. This underground water, which is properly called GROUND WATER, gathers in the pores and cracks of rocks or flows through underground passageways. The flow of ground water depends very much on the character of the rocks. Some rocks are porous (or sponge-like), containing many air-filled pores when dry. Such rocks soak up water. Other rocks are permeable, that is, they are solid, but water can seep through joints and horizontal bedding planes. However, some rocks are impermeable and no water can pass through them.

Areas of rock in which all the pores are filled with water are said to be part of the zone of saturation. The top of this zone is called the WATER TABLE. In dry areas, the water table may be far from the surface whereas, in wet areas, it may be just below the surface.

Porous and permeable rocks can form an AQUIFER – an underground reservoir of water. Wells dug down to the aquifer will yield water. Sometimes, an aquifer is trapped between layers of impermeable rocks, through which water cannot escape. Such water-bearing layers are called confined aquifers. Artesian wells are special kinds of wells which tap the water in confined aquifers. However, at least one end of the aquifer must be on the surface, so that rainwater can enter and replenish the water supply. In artesian wells, the water is usually under pressure because the confined aquifers are tilted. Hence, when a hole is drilled down to the aquifer, the water gushes upwards. Examples of artesian wells occur in the London basin and the GREAT ARTESIAN BASIN of Australia.

Youthful rivers

The youthful stage of a river is usually characterized by a steep gradient, or slope. Mountain streams flow so quickly downhill that they have the force to move boulders. Loose bits of rock, broken off by weathering, tumble down into

Above: The Iguaçu Falls, on the border between Argentina and Brazil, plunge down more than 70 metres over a rocky ledge which is over 3 km wide. The falls are a great tourist attraction.

Below: The gorge of the River Zambezi below Victoria Falls, forms part of the boundary between Rhodesia (or Zimbabwe) on the left, and Zambia on the right. The River Zambezi,

which was explored by the Scottish missionary David Livingstone in the 1850s, flows into the large, man-made Lake Kariba.

begin to develop when the head of a mountain valley is deepened by the action of frost-shattering and ice. As snow accumulates, it becomes compacted into ice. As the ice moves, the back and bottom of the cirque are gouged away. The sharp ridge separating 2 cirques is called an arête. After the ice has gone, the armchair-shaped cirques often contain lakes called tarns.

Clints are blocks of limestone on the surface. They are formed when rock JOINTS are enlarged by solution weathering. The widened joints are called GRIKES.

Cockpit Country, in Jamaica, is an example of KARST scenery. It is a wild and uninhabited area composed of innumerable conical hills, dividing circular pits often over 150 metres deep.

Crag-and-tails are formed when a glacier passes over or around a hard, igneous rock outcrop. The ice removes most of the loose debris, forming a steep face, but drops material behind the rock, thus creating a gentle slope or 'tail'.

Crevasses are cracks in the ice in a glacier. They are

Crevasses

caused by the movement of the glacier, particularly when it moves over an uneven valley floor. The cracks may be hundreds of metres long. They can be dangerous to climbers because they are often hidden.

D **Deflation** is the transport of loose surface debris, sand and dust, by the wind.

Deltas are deposits of sediment formed when a river enters the relatively still waters of a lake or sea and drops its load. Normally, there is little or no current in

the water. Otherwise, the sediments would be swept away. Deltas are often triangular in shape, like △ (delta), the fourth letter of the Greek alphabet, after which they were named. Deltas vary in size from small deltas in lakes to the huge Ganges delta in India and Bangladesh. Arcuate deltas, such as the Nile delta, are formed in a broad arc-shape. Bird's-foot deltas, such as the Mississippi delta, contain long, thin fingers of sediment that stretch out to sea.

Denudation is the name for the general wearing down of

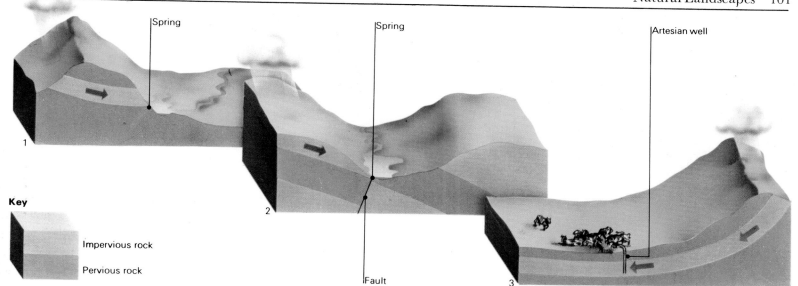

Key

Impervious rock

Pervious rock

mountain streams. The faster the stream flows the larger the fragments it can carry. As the water pushes the fragments along, they rub against the bed and sides of the stream, eating further into the land surface. This process is called abrasion. Several other kinds of erosion also operate. Attrition occurs as the loose fragments rub against each other and break into smaller and smaller pieces. Hydraulic erosion is caused by rushing water forcing air and water into cracks in rocks, with the resulting explosive effect. Finally, solution erosion occurs as the water dissolves minerals in the rocks.

Because most of the erosion is downwards, the VALLEYS develop steep sides and a characteristic V-shape profile. In arid areas, where there is little weathering on the valley sides, the stream or river may cut down deeply into the land, wearing out a deep GORGE or CANYON.

Rapids and waterfalls

Rapids and waterfalls often form in the youthful stage. Soft rock is worn away faster than hard rock. Therefore, soft rocks are eroded to a lower level more quickly and a 'step' may be formed in the river's course. Because the gradient increases at this point, the speed of the river also increases and rapids are formed. But as the softer rock continues to be eroded, the step becomes larger and a waterfall evolves. The erosive power of a waterfall is such that it gradually recedes upstream, leaving a gorge downstream.

Above: Springs occur where pervious, water-bearing rocks, called aquifers, reach the surface (1). They are sometimes created by faults (2). Artesian wells (3) tap tilted aquifers from which water rises under pressure.
Below: Continuous and torrential downpours of rain can often cause rivers to overflow their banks and flood surrounding areas. This photograph shows the disruption that can be caused if flooding occurs in an urban area.

the land by weathering, erosion and transportation.
Deposition is the laying down of eroded sediments which have been transported by water, ice or winds.
Deserts are dry areas, usually having less than 250 mm of rain per year. As a result, they have little vegetation, animal life or population. Types of deserts include the cold deserts of the tundra region around the North Pole; the temperate continental deserts, far from the sea such as the Gobi desert of central Asia; and the hot deserts of tropical areas, such as the SAHARA. The Sahara is the sunniest place on Earth and the world's highest temperature in the shade, 57.7°C, was recorded at Al'Aziziya, in the Libyan part of the Sahara. In the Atacama desert, near Calama in Chile, no rain has been recorded for 400 years. There are 3 main types of desert scenery: hammada, with its bare, polished rock surface; reg, with its angular pebbles and stones; and erg, which is sandy desert. Because there is so little water the main agent of erosion is the wind.

Death Valley, USA

Drainage patterns

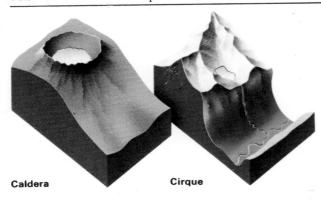

Caldera   Cirque

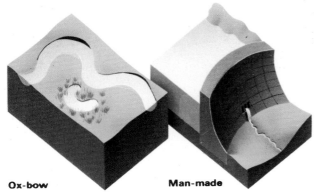

Ox-bow   Man-made

Left: The diagrams show 4 types of lakes. Some form in calderas, which are basin-shaped volcanic craters (*top left*). Some lakes, called tarns, occupy the bottoms of armchair-shaped mountain cirques (*top right*) which were worn out by ice. Ox-bow lakes (*bottom left*) are formed when rivers straighten their courses and leave behind an abandoned meander as a lake. Other lakes, (*bottom right*) are man-made. The dams are usually built to provide irrigation water or a head of water which will power nearby hydro-electric power stations.

Below: This salt lake is about 300 km south-west of Alice Springs, in the desert heart of central Australia. The lake is often dry, because the Sun evaporates all the water, leaving a flat coating of salt. In the background is Mount Connor, which is a mesa – a table-like upland whose top layer has resisted erosion.

Mature rivers

The mature stage begins when rivers leave the mountains. The decrease in gradient slows down the river. As a result, it drops much of the debris collected in its youthful stage, except when it is in flood, when the increased volume of water compensates for the lack of slope. Mature rivers are still vigorous, but lateral (sideways) erosion becomes more important than downwards erosion. Hence, the river widens its valley. Valley widening increases when the river develops bends, called MEANDERS. Meanders are even more marked in the old age stage, where the river gradient is extremely slight.

Rivers often erode through the necks of meanders, straightening their courses. The meander is then left as an OX-BOW LAKE, or abandoned meander. The river's water may also be diverted by river capture, or blockage by glacial material. Changes of course by rivers often cause severe flooding.

Rivers in old age

The last part of the river's journey to the sea, the old age stage, extends over a wide floodplain, which is often so flat that the rivers easily overflow their banks. In some areas, silt and other material piles up on the sides of the river forming natural banks, called *levées*. Often, floods spread vast amounts of fine silt over the surrounding plain. This silt is often very fertile. For example, the silt deposited by floods in the Nile valley supported the civilization of ancient Egypt.

As the sluggish, meandering river flows into the sea or a lake, suspended sediments fall to the bottom, sometimes building up to form a DELTA. Deltas occur when the offshore currents are not strong enough to wash away the sediment, which piles up in a group of mudbanks breaking up the river's path. The two main kinds of deltas are the arcuate (fan-shaped), such as the Nile delta, and the bird's foot delta of the Mississippi River in the United States. The amount of material carried into the sea by many rivers is tremendous. It has been estimated that the Mississippi River transports 700 million tonnes of material into the sea every year.

In moist regions, over millions of years, rivers can wear down the land almost to a level plain. This end result is a peneplain (which means

---

Differential erosion occurs because soft rocks are eaten away at a faster rate than harder, more resistant types.

Divide, see WATERSHED.

Drainage pattern is how the arrangement of a river system looks from the air on a map. It may be dendritic – like a tree with irregular branches; trellis – like a network of parallel streams joined together by cross branches; or radial – starting off from a single high point with the streams flowing away from each other. The main factors influencing

drainage patterns are the slope of the area, the types of rocks and their structures, the existence of volcanic activity, the geological history of the area and the climate.

Dreikanter, see VENTIFACT.

Drumlins are elongated, egg-shaped mounds of boulder clay, up to 2 km long and 60 metres high. They often occur in swarms, creating the so-called 'basket of eggs' scenery. Geologists think that they were formed from material dropped by a slowly-moving ice sheet.

Dry valley is a valley in which no water flows. The

original river that carved the valley must have been diverted by a glacier, captured by another drainage system, or it may have ceased to flow because of a decrease in the water supply and a lowering of the WATER TABLE.

Dunes are heaped masses of loose, unconsolidated sand. They are formed when the winds blow steadily in one direction. The main types are the barchans, which are crescent-shaped and more than 200 metres long; and seif dunes which are long narrow ridges of sand up to 50 km long. Sand

dunes also occur along some coastlines, such as the Landes in France. The world's highest dunes (430 metres) are in the sea sand of Issouane-n-Tiferine in the Algerian Sahara.

E **Earth pillars** are shaped like giant mushrooms with clay stalks and rock heads. The rock protects the ground below and so it survives. In Austria, examples up to 60 metres high have been found.

Engineering geology is the use of geology in the construction industry. It is con-

cerned with finding raw materials for building and the study of the geological character of the building area, including the strength

Zeugen

Above: Ayers Rock, in central Australia, is a sandstone inselberg, which means 'island mountain'. Inselbergs are the remains of land surfaces which have mostly been eroded away. This impressive natural feature is over 2.5 km long and 1.6 km wide. It rises about 360 metres above the surrounding plain.

Right: The Amazon River, shown here near the river port of Manáus, Brazil, stretches as far as the eye can see. Although at this point it is still over 1,000 km from the sea, it has already formed an extensive flood plain.

'almost a plain'). Sometimes, MONADNOCKS (isolated hills of resistant rock) survive to break the landscape's monotony. River erosion is at an end in such areas. However, the land may later be uplifted, creating another slope and the river is then said to be rejuvenated. Rejuvenation causes erosion to start all over again.

Lakes

Lakes are large sheets of inland water. They may be the sources of rivers or they may occur along the course of a river if the river valley is blocked by landslides, man-made dams, glacial material, lava, and so on. Other lakes form in volcanic craters, but the greatest number of lakes are formed by the erosive action of glaciers and ice sheets (*see pages 106-107*).

Landslides and floods

Landslides occur in mountain areas when rocks or other material on slopes are loosened and lubricated by ice or water, or moved by earthquakes. Gravity does the rest. Soil on slopes may also 'creep' downhill and collect in valleys. Any of these events can cause a flood.

The world of limestone

When water travels underground, it can erode and redeposit rock in the same way as surface water. The main difference is that groundwater erodes largely by chemical erosion and the material is transported in solution. Limestone consists mostly of calcium carbonate, which is not soluble in pure water. However, if the water contains carbon dioxide dissolved from the air or from organic material, it becomes a weak carbonic acid. This acid reacts chemically with limestone, turning it into calcium bicarbonate, which is readily soluble.

Limestone is riven by JOINTS (vertical cracks) and BEDDING PLANES (horizontal cracks), which allow water to percolate easily through it. On the surface, some cracks are enlarged by chemical erosion into funnel-shaped sink or SWALLOW HOLES. Streams sometimes enter these holes and plummet downwards. The water may then travel hundreds of kilometres under the ground, dissolving away vast caverns. Sometimes, so much limestone is removed that the roof of a cavern collapses. If the cavern is near the surface, a gorge is formed. Eventually, however, the underground water collects into streams and re-emerges at the base of the limestone. Some caves are extremely deep. The deepest-known cave is in the Pyrenees in France, where a depth of more than 1,330 metres has been measured.

Below: The diagram shows some features which make karst (limestone) scenery so distinctive. The surface is pitted by grikes (dissolved joints) which separate the clints (blocks of rock). Water drains through the joints or into swallow holes or pot holes. Pot-holers explore limestone caves, which may be festooned with stalactites and stalagmites. They trace underground streams until they surface at places called resurgences.

Karst landscapes

Most limestone landscapes have a distinctive surface character, called Karst topography, after the carst district in the Dinaric Alps of Yugoslavia. Because water percolates so readily into limestone, the surface is often dry and plants grow only in hollows where clay accumulates. Bare limestone surfaces are divided into blocks, called clints, separated by eroded joints, called grikes. Karst scenery occurs in the COCKPIT COUNTRY of Jamaica and in the North American Kentucky plateau, which has more than 60,000 sink holes, hundreds of caves and 231 kilometres of continuous underground passageways in the MAMMOTH CAVE NATIONAL PARK.

Limestone caves

An eerie world made entirely of calcium carbonate is formed in some caves. Drops of water,

|Stalagmite |Swallow |Grikes |Pot-holer |Clints |Roof fall |Syphon |Underground stream |Resurgence

gouged out deep valleys. Sogne Fiord in N

Left: Clints and grikes can be clearly seen in this photograph of a limestone pavement above Malham Cove, Yorkshire, England.

Below: This impressive photograph of a mountain skyline was taken in the Dolomites, an Alpine region in the Austrian Tyrol and north-eastern Italy. The mountains are named after the mineral dolomite, or magnesian limestone. Like other limestones, dolomite also produces karst scenery. The Dolomites are a popular tourist attraction.

highly charged with calcium carbonate, often hang for some time on a cave roof before falling. While they hang, some of the water is evaporated and a film of calcium carbonate is deposited on the roof. If the water always drops from the same place, successive films grow into rock icicles, or STALACTITES. When the water drops onto the floor, similar deposits form in upward-growing STALAGMITES. Eventually, they may join and form pillars of calcium carbonate. The speed of growth varies. Some take 2,000 years to grow only one centimetre. But a stalactite in Ingleborough Cave, in Yorkshire, England, grew by 7.6 centimetres in only ten years.

Various other strange and beautiful deposits form in some caves. Fringed curtains of calcium carbonate form when water drips down from an irregular crack in the roof of a cave. Water flowing across a cave floor may deposit flowstones. There are also various delicate deposits, some of which resemble flowers.

Limestone caves offer a challenge to explorers, called pot-holers. Exploring caves can, however, be dangerous, especially when heavy rain raises the water level in caves.

area by erosion and consequent deposition. There are 2 main types of glacier. Valley, alpine or mountain glaciers start in a CIRQUE and move down a valley. The size of these glaciers varies from a few metres to about 50 km. Examples occur in the European Alps and on South Island, New Zealand. Piedmont glaciers are 2 or more valley glaciers which have combined as they move onto flatter land below the valleys. An example is the Malaspina glacier in Alaska, which covers about 4,000 km². *See* ICE SHEETS.

Gorge is a deep, narrow, vertically-sided valley. The term canyon is usually used for large gorges.

Mud flow after heavy rain

Grand Canyon, in Arizona, USA, is one of the natural wonders of the world. It is between 3 and 39 km wide and reaches a depth of 1,700 metres in places. In the canyon, the horizontal rock layers are exposed. At the bottom, the river flows through rocks of Pre-Cambrian age, more than 1,750 million years old. At the top, there are Pleistocene deposits, about 10,000 years old. The Colorado plateau was once a flat plain. It was then uplifted and the river was rejuvenated, giving it the power to cut the canyon.

Gravity transport is the movement by gravity of weathered debris. This may

Coastal erosion in Taiwan

be down a slope, in the form of SOIL CREEP, mud flows or AVALANCHES; or by mass movement along a place of

Ice erosion

About 2.15 per cent of the world's water is frozen into ice. Much of this ice is locked in the great ICE SHEETS of Antarctica and Greenland, although smaller ice caps and many mountain GLACIERS occur in other areas. Bodies of ice, which move down slopes under the force of gravity, have changed the face of parts of the Earth. During past ice ages, ice covered as much as 30 per cent of the Earth's land surface. Today, however, it covers only about 10 per cent of the land.

Some areas are permanently covered with snow. In polar regions, the snow-line, the lowest level at which snow remains unmelted all the year round, is at sea level. Moving towards the equator, however, permanent snowfields are restricted to higher and higher areas. The snowline is at 1,500 metres in Norway, 3,000

Below: The Aletsch Glacier, in the Bernese Alps of Switzerland, is about 26 km long. This great river of ice moves slowly downhill, transporting weathering rock fragments (moraine) within it and on its surface. The action of ice and weathering (especially frost shattering) have moulded the landscape, which exhibits many of the features that characterize a glaciated upland, such as hanging valleys and knife-edged arêtes (ridges between the armchair-shaped cirques).

metres in the European Alps, 5,000 to 6,000 metres in the Himalayas, and at more than 6,000 metres in central Africa and in the Andes.

Snowfields on mountains often form in semi-circular hollows called CIRQUES (a French term), corries (in England and Scotland) or cwms (in Wales). Here, the snow is compacted into ice, called névé or firn. This ice eventually becomes glacier ice, which spills over the lip of a cirque and flows downhill in a valley glacier.

The path of a glacier

The speed at which glaciers flow is usually a metre or so a day. The speed depends on factors such as the size of the glacier and the steepness of the slope. The record speed of 60 metres per day occurred for a short period at the Black Rapids Glacier, Alaska, in 1936–37. This speed was

weakness, such as a fault or a joint.
Great Artesian basin, in east-central Australia, is a dry region. But beneath the basin is a porous AQUIFER, which is supplied with water because it outcrops in the Great Dividing Range in eastern Australia. The artesian wells in the basin make possible Australia's important stock-rearing industry.
Greenland ice sheet covers 1.74 million km². The average thickness of the ice is 1.6 km, but it reaches a maximum thickness of about 3 km at the centre.

Grikes are the widened joints which separate CLINTS on limestone surfaces.
Ground water fills cavities, openings and spaces in the rocks beneath the surface. Ground water occurs between the grains of porous rocks, or in the joints and cracks of pervious or permeable rocks. The main kinds of ground water are *meteoric*, from rainwater which soaks into the rocks; *connate*, from water which was trapped when the sedimentary rocks were first laid down; and *juvenile*, released from underground magma.

H Hanging valleys are tributary valleys which were not glaciated to the same extent as the main valley to which they are joined. They are, therefore, left high above the main, over-deepened U-shaped valley. Waterfalls often occur at the junction.

View down a glaciated hanging valley

Hardpan is a layer of strongly-cemented material found, in places, just below the surface of the ground. It is formed, in hot, dry areas, when water is drawn towards the surface and the minerals contained in the water are deposited. Hardpans can be a major obstacle to farming, because they are difficult to break up by ploughing.
Humus is dark, organic (once-living) matter found in soils. It consists of the remains of decaying plants and other organisms. It is important as a source of

probably attained because of an earthquake which shook a tremendous amount of snow onto the glacier's source.

Once on the move, glaciers are effective eroding, transporting and depositing agents. They pluck away pieces of rock which are then frozen into their bottoms and sides. The rock fragments scratch, polish and grind the bedrock below the glacier. Valleys are gouged out to form U-shaped profiles, with flat bottoms and steep sides.

Hard rock outcrops are smoothed and rounded. They survive as ROCHES MOUTONNÉES or CRAG-AND-TAILS. The uneven floor of the valleys causes glaciers to crack, forming CREVASSES – dangerous, gaping holes in the ice, into which weathered rocks tumble. The surface of the ice is also littered with loose rocks called MORAINE.

Above: The photograph shows ice caves in a Norwegian glacier. Most ice caves are formed in the snouts of glaciers by streams which flow out from the base.

Right: The diagram shows various features associated with glaciated scenery. The glacier, riven by deep crevasses, has over-deepened its valley, leaving tributary valleys 'hanging' at a higher level. Waterfalls often plummet down from hanging valleys. Another feature of glacial erosion is the crag and tail. This occurs when the ice flows over a particularly hard mass of rock, such as a volcanic plug. The upstream side is steep, but the downstream side has a gentler slope, because it has been filled in by moraine. Ridge-like terminal moraines mark the ends of glaciers. If a series of them occur, they mark the successive stages of a glacier's retreat. Beyond the terminal moraine, there are often fluvio-glacial deposits – that is, moraine which is spread by streams flowing from the glacier or from a meltwater lake dammed behind the terminal moraine. One such feature is the alluvial fan. Drumlins are low, oval-shaped hills of boulder clay.

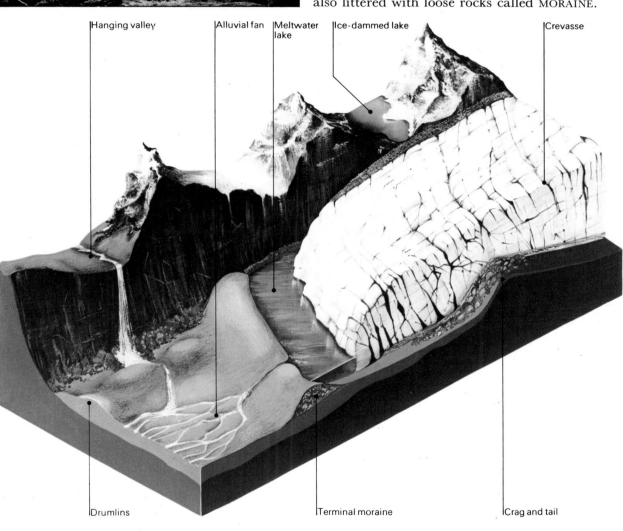

Hanging valley | Alluvial fan | Meltwater lake | Ice-dammed lake | Crevasse

Drumlins | Terminal moraine | Crag and tail

plant food and it helps to bind the soil together.

Ice caps are small ICE SHEETS. They occur in parts of northern Canada, Iceland, Norway and Spitzbergen.
Ice sheets, sometimes called continental glaciers, cover vast areas and contain great thicknesses of ice. The world's largest ice sheets cover most of ANTARCTICA and GREENLAND.
Inselberg is a German word, meaning 'island mountain'. These steep-sided, usually round-topped rock masses rise above a flat

PEDIMENT. They are made of resistant rocks, such as granite, and are particularly common in dry areas in Africa and Australia.

Weathered limestone

Juvenile water is 'new' water which has been released from magma. It is, therefore, the only GROUND-WATER that has not come from the sea or the atmosphere.

Kames are mounds of sand and gravel formed from sediment which was deposited by a river flowing from a melting glacier. They occur in OUTWASH PLAINS.
Kaolin, see CHINA CLAY.
Karst landscape is the name for weathered limestone scenery, after the Karst district of Yugoslavia.

Kettle-hole is a depression in the land, formed when a buried block of ice from a glacier eventually melts.

Lacustrine deposits are sediments which accumulate in lakes.
Lakes are inland bodies of water. More are caused by glaciation than by any other agency. The largest are as follows:

Lake	Area (km²)	Location
Caspian	393,898	USSR
Superior	82,414	America
Victoria	69,485	E. Africa
Aral	68,682	USSR
Huron	59,596	USA

Wastwater, England

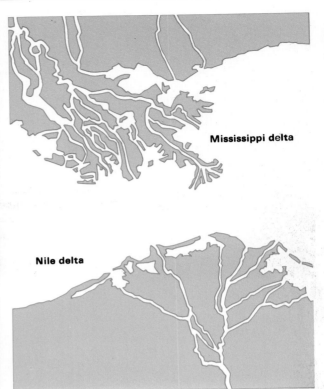

Mississippi delta

Nile delta

Glacial deposition

At the snout (end) of a glacier, melting water flows outwards in narrow streams. These streams deposit fine sand and silt from the ice which form long ridges, called ESKERS. Glaciers also drop a vast amount of moraine when they melt. This material often piles up in great ridges. There are various kinds of moraine, depending on the part of the glacier in which it was formed. Till, or BOULDER CLAY, consists of unsorted deposits of clay and rocks laid down directly by melting bodies of ice.

Moraine may sometimes block the path of a river and create long, narrow stretches of water, called ribbon lakes. Also, blocks of ice buried in boulder clay melt. The clay slumps down, forming a KETTLE-HOLE, which may later become a lake. Other glacial lakes form in cirques.

After an ice age has ended, the land is unrecognizable. The shapes of valleys and the routes of rivers are changed and the formation of new ridges all give the land a new face. For example, Finland's scenery changed during the Pleistocene Ice Age, when the ice gouged out many hollows which now make it a land of lakes.

Above left: Chesil Beach links the mainland of Dorset to the Isle of Portland, in southern England. This shingle beach is a type of wave-deposited spit, or bar, called a tombolo, after similar features in Italy.
Above right: The diagram shows 2 kinds of deltas. The Mississippi delta is a lobate, or segmented, delta. It is usually called a bird's foot delta, because of its shape. The Nile delta resembles an inverted Greek letter, Δ (delta). But the seaward side is curved or arched, like a fan. Hence, such deltas are named arcuate deltas.

The sea

The oceans contain most of the world's water and have a great influence on climate. In addition, around coastlines, the restless sea is continually acting as an agent of erosion, transportation and deposition.

Some coastlines, called coasts of submergence, have been drowned by the sea. For example, this may happen at the end of an ice age, when the sea level rises. River valleys are then drowned to become rias and glaciated valleys become FIORDS. Other coastlines have been uplifted by Earth movements, forming coasts of emergence. Along such coasts, cliffs and 'raised' beaches are sometimes found far inland.

The sea is never still and waves, currents and tides can break down even the most resistant rocks. Coastal erosion takes several forms. Hydraulic action occurs when seawater drives and compresses air into cracks and crevices in rocks. When the air escapes, the release of the pressure is explosive, shattering the rock. Waves hurl rocks and other debris against the land in a process called ABRASION. Also, some coastal rocks, such as chalk, are moderately soluble in

---

Lambert Glacier, in the Australian Antarctic Territory, is the world's largest. Discovered in 1956, it is 402 km long and, with the Fisher Glacier, it has a total length of 514 km.
Landslide, see GRAVITY TRANSPORT.
Laterites are red soils formed in humid tropical regions. Heavy rain leaches the soil (see LEACHING), but insoluble substances (iron, bauxite and manganese) are left in the top layer.
Leaching is the downward movement of material dissolved in water, from the top

layer of soil. The material may accumulate lower down in a HARDPAN.
Limestone pavements are outcrops of limestone on the surface. They are often divided into CLINTS, which are separated by GRIKES.
Load is the material carried by an eroding agent, such as a river, glacier or wind.
Loess is an accumulation of wind-blown yellow dust (mainly silica). It is often transported thousands of kilometres before being deposited.
Longshore drift transports debris along a seashore. The

movement is a zig-zag one, because the waves hit the shore at an angle, throwing debris up the beach. But then the water and debris

Meanders

are sucked back into the sea at right angles to the shore.

M **Mammoth Cave National Park,** in Kentucky, USA, contains the world's largest known cave system. More than 60,000 SWALLOW HOLES are found in the LIMESTONE PAVEMENT. The passageways total 231 km.
Meander is a natural bend in a river or stream, especially common in floodplains. On the outside of a meander, the water runs faster and so it erodes deeper. On the inside of the bend, the water slows down and deposits

some of its load. Meanders are thus enlarged. Sometimes, rivers cut across the necks of meanders to form OX-BOW LAKES.
Mer de Glace Glacier, in the Mont Blanc massif of the Alps, is 13 km long, 2 km wide and 150 metres thick.
Mesas are flat-topped hills formed when a hard rock layer resists the erosion which has flattened surrounding areas. Buttes are small mesas.
Millet-seed sands are the worn down grains of sand formed by rubbing against obstacles and each other.

seawater and they are slowly dissolved. The power of the sea has been well illustrated by storm waves which have shifted concrete blocks weighing more than 1,000 tonnes. Thus, waves are the most destructive weapons in the sea's armoury.

The sea has carved the many familiar shapes of coastal scenery. Cliffs, sometimes hundreds of metres high, are formed as the sea cuts away at the base, causing slabs of overhanging rock to collapse. The longest stretch of cliffs is the 450-kilometre-long Saw Tooth Cliff fringing the Great Australian Bight. The flat area at the base of a cliff is a wave-cut platform.

Of course, rocks erode at different rates.

Above left: Hong Kong is a small and crowded territory. Some areas, such as this industrial area, have been reclaimed from the sea.
Above right: The chalk cliffs at Studland in Dorset, in southern England, are being eroded by the waves. The offshore stack was once part of the mainland.
Below: The wavelength is the distance between 2 crests. The wave height is the vertical distance between a crest and a trough. When a wave passes through the sea, the water particles move around in a circle.

Headlands are formed from resistant rocks, while bays are eroded from weaker ones. Sea caves are hollows cut into cliffs by wave action. If two caves on either side of a narrow headland meet, a sea arch is formed. Continued erosion may cause the roof of the arch to collapse. A pillar of rock, called a stack, is thus detached from the land and left standing in the sea.

Transport and deposition

Waves and currents are active in carrying eroded material and dumping it elsewhere. As the material is transported, it is often eroded further, rounded and sorted according to size. LONG-SHORE DRIFT is the name given to the way in which debris is moved along a coast.

When an obstacle, such as a headland, occurs, part of the load of debris is dropped. The deposits pile up in a narrow ridge called a SPIT. Spits extend due to longshore drift; currents may cause them to curve in towards the shore to form a hook. Eventually, the spit may link up with land opposite the headland. The sea is then cut off and the spit (now called a bar) encloses a lagoon.

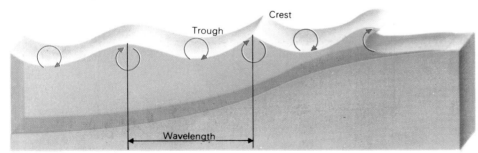

Crest
Trough
Wavelength

Monadnocks are isolated hills on a PENEPLAIN. They survive because they are composed of very resistant rock. They are named after Mt. Monadnock in the USA.
Moraines are ridges or heaps of boulders, sand and gravel, or clay deposited by a glacier. Terminal (or end) moraine is piled up at the front of stationary glacier. Rocks on the sides of a glacier are called lateral moraine. When two glaciers meet, the lateral moraines join to become a medial moraine. Debris frozen in the ice is called englacial

moraine. Sub-glacial moraine consists of rocks at the base of a glacier.
Mushroom rocks, see ZEUGEN; EARTH PILLAR.

N **Névé,** or firn, is snow which has been compacted and frozen into ice. It compacts further to form glacier ice.

Niagara Falls

Niagara Falls, on the USA-Canada border, lie between Lake Erie and Lake Ontario. There are 2 main falls: the Horseshoe Falls in Canada and the American Falls in the USA. The Horseshoe Falls are slightly higher than the American Falls at 57 metres.
Nile River is the world's longest. It stretches 6,690 km from the East African plateau to the Mediterranean.
Nunatak is an Eskimo name for a rock mass that juts through an ice sheet.
Nutrient is an element or chemical food required by growing plants.

O **Oases** are places in deserts where water occurs at the surface. They range from small waterholes to river valleys, such as that of the Nile.
Outwash plain is an area covered by sediment deposited by melting glaciers.
Ox-bow lakes are formed when a river cuts through the neck of a meander, straightening its course. The abandoned meander forms a lake, which becomes a marsh before it finally dries up. Other names include mortlake, billabong and bayou.

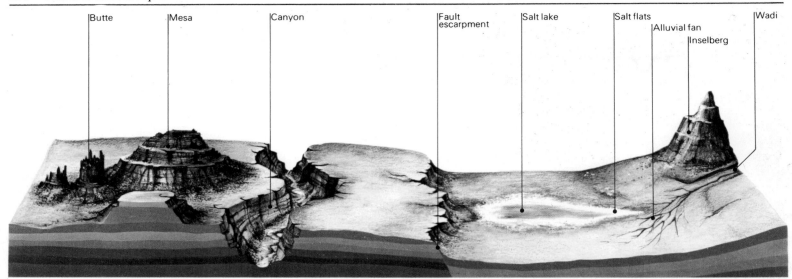

Butte | Mesa | Canyon | Fault escarpment | Salt lake | Salt flats | Alluvial fan | Inselberg | Wadi

Wind erosion

In arid areas, wind is the chief agent of erosion, transportation and deposition. The power of the wind to erode the land is seen when a drought occurs in a farming region. The soil dries up and the vegetation cover is reduced. As a result, the soil is turned into dust. For example, large areas of once-rich farmland in the Mid-West of the USA were reduced to dust bowls as the wind stripped away the rich topsoil. The action of the wind is called deflation.

DESERTS are empty places with a sparse plant and animal life. Although there are cold deserts, for example the Gobi in Central Asia, hot deserts are the first to spring to mind, including the North African SAHARA, the world's largest, and the Atacama in South America. In general, the average annual rainfall in hot deserts is less than 250 millimetres and the daytime temperatures are high (40°C in the summer). These are ideal conditions for wind erosion, because there is no surface water or vegetation to bind the rock and sandy soil together to resist the wind's continuous attack.

Wind-blown, hard sand grains act like abrasives, cutting, pitting and polishing rocks. Because sand grains are heavy, they are seldom lifted more than two metres above the ground, so wind-powered sand-blasting occurs mostly at low levels. For this reason, most people find it uncomfortable to wear shorts when the wind starts to blow in deserts. Desert-dwellers therefore traditionally wear long robes.

Above: This arid landscape is like those used in American Western films. The plateau (*left*) which has been uplifted along a fault, is broken by mesas and buttes, remnants of an earlier surface. Canyons occur, because weathering of valley sides is minimal in warm, dry areas, while down-cutting by rivers is fast. On the right is a wadi cut by flood water during freak storms. Flood water deposits alluvial fans and may form salt lakes.

Wind-blown sand scours the land, carving out deep depressions in the surface rocks. The wind also creates strange shapes in rocks. The bases of large rocks are worn away, while the tops are untouched. As a result, large rocks are carved into mushroom shapes, with narrow, eroded stems supporting large, uneroded tops. Sometimes, rocks containing a number of layers are carved into jagged-edged pedestals, called yardangs. A large resistant body of rock rising above the desert debris is called an INSELBERG (German for 'island mountain'). A famous inselberg is AYERS ROCK in central Australia.

Pebbles and boulders are also polished and grooved by wind action to form VENTIFACTS, which often have smooth sides separated by sharp edges. Even the grains of sand are themselves rounded and worn down. The rounded, small grains are known as millet-seed sands.

Desert landscapes

There are three main kinds of desert landscapes: hammada, reg and erg – three terms derived from Arabic. Hammada consists of desolate areas of bare rock. Reg is a large, flat area strewn with gravel and small stones. However, the erg, or sandy desert, is the best-known desert landscape. It covers about 20 per cent of all the deserts in the world.

One of the chief features of erg landscapes is the sand DUNE. These mounds of sand are deposited and shaped by the wind. They also occur in sandy coastal areas. One type of dune,

P **Pediment** is a flat plain of eroded rock, sometimes covered by alluvium, in an arid region.
Pedology is the study of soil.
Peneplain, meaning 'almost a plain', is a flat surface which occurs when an area has been eroded down to BASE LEVEL.
Permeability of a rock is a measure of how easily liquid (usually water) can pass through it.
Podsol is a soil found in cool wet areas. Greyish-white in colour, it is heavily leached (*see* LEACHING).

Porosity of a rock depends on the gaps between the grains which allow water to be absorbed. Porous rocks may be permeable (letting water pass through them) or impermeable, such as clay.
Pot-hole is another name for swallow hole. Pot-holing is a name for the exploration of underground caves.
Prairie soils are dark brown soils found in middle latitudes. They are coloured by HUMUS formed from decayed grass.

R **Rivers.** The longest rivers are the Nile (6,690

Waterfall in Yosemite Park, USA

km), the Amazon (6,570 km), Mississippi-Missouri (5,690 km), Ob-Irtysh (5,570 km), and the Yangtse (5,520 km).

Roches moutonnées are rock projections from a valley floor which have been smoothed and rounded by the advance of a glacier. The upstream side is smooth, but the downstream side is rough.
Run-off is the water that flows over the land's surface.

S **Sahara**, in North Africa, is the world's largest desert. It covers about 8.4 million km².
Salto Angel, a waterfall in Venezuela, is the world's highest. It has a total drop of

979 metres, with the largest single drop of 807 metres.
Salto dos Sete Quedas, a waterfall on the Brazil-Paraguay border, has the world's greatest flow of water – over 500,000 cubic metres per second.
Scree, or talus, is an accumulation of rock debris formed by weathering.
Sea caves are hollows in sea cliffs eroded by wave action. If 2 caves cut in a headland meet, a sea arch is formed. When the arch collapses, the seaward end becomes an isolated rock, called a stack.

the barchan, is crescent-shaped, with two long, curved arms pointing in the direction of the prevailing wind. Another type, the seif (Arabic for 'sword') dune, is a long ridge of sand which develops parallel to the wind direction. The smallest dunes are merely ripple marks, but some are hundreds of metres high. They can also move at about 10 metres a year.

Deserts are seldom completely dry, because occasional thunderstorms cause flash floods. Torrents sweep along gorge-like channels called WADIS and temporary lakes, or shotts, are formed in the mountains. These torrents have tremendous erosive power. After such floods, dormant plants spring to life, often scattering their seeds only two weeks after flowering. The desert then becomes, briefly, a blaze of colour.

Right: Seif dunes are long, narrow ridges of sand which lie parallel to the direction of the prevailing wind. On the other hand, barchans are crescent-shaped dunes, which lie at right angles to the prevailing wind.

Below: This magnificent landscape is called the 'Moab deadhouse'. It is in south-eastern Utah in the USA. The valley forms part of the spectacular Grand Canyon, which was carved into the Colorado plateau by the Colorado River. The Colorado plateau was formerly a level coastal plain but Earth movements raised the plain to form a plateau.

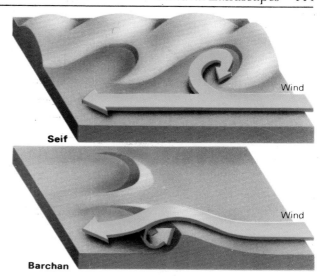

Seif

Barchan

Wind

Wind

Durdle Door, Dorset

Sink hole, see SWALLOW HOLE.
Snow-line is the lowest level at which snow remains unmelted all the year round.

The height of the snow-line depends on the latitude.
Soil is an accumulation of loose, weathered rock in which plants grow. *See also* HUMUS.
Soil creep, or solifluxion, is the slow, downhill movement of soil due to gravity.
Speleology is the study of underground tunnels and caves.
Spits are long, narrow strips of debris deposited by the sea. Sometimes, spits extend from one headland to another. They are then called bars and enclose, on the landward side, a lagoon.

Stack, see SEA CAVES.
Stalactites are deposits, composed mostly of calcium carbonate, which grow downwards from the roof of a tunnel or cave. The largest is 59 metres long in the Cueva de Nerja, Spain.
Stalagmites are deposits like stalactites, but stalagmites grow upwards from a cave floor. The highest is 29 metres in the Aven Armand cave, Lozère, France.
Sugar Loaf Mountain is a 395-metre-high granite INSELBERG in Rio de Janeiro, Brazil.
Swallow or **sink holes** are funnel-shaped openings in

LIMESTONE PAVEMENTS. They link the surface with underground passageways and caves.

Tasman Glacier, on the eastern side of Mt. Tasman, New Zealand, is almost 30 km long.
Till, see BOULDER CLAY.
Tombolo is a spit or bar which links an island to the mainland or to another island.
Topography is a term used for the physical features of the land's surface.
Tributary is a river which flows into a larger one.

U-shaped valleys are over-deepened, steep-sided valleys formed by glaciation.

Landscape in Provence

Loess deposits

A special kind of aeolian (wind-blown) deposit is called LOESS. Loess is a yellowish, fine-grained material made up of dust which usually originates in deserts. A vast belt of loess stretches from Germany and France, where it is thin and glacial in origin, across the USSR to China, where it is derived from desert erosion. The loess layers in China are extremely thick. The Hwang Ho is coloured yellow by the loess which it transports.

Deserts on the march

Many desert features are clearly the result of water erosion, which occurred when the areas had moist climates. This is true of the Sahara where, thousands of years ago, rich grasslands supported nomadic communities and an abundance of savanna animals. This is known from the ancient rock paintings found in the Sahara.

Climatic changes are continuing. In some areas, deserts appear to be spreading. For example, the Sahel, the dry savanna zone south of the Sahara, suffered prolonged droughts in the 1960s and 1970s. Water holes dried up, vast numbers of cattle died and many people starved. The long droughts killed off much of the plant life and the wind removed the precious topsoil.

Oases

Most human settlement in deserts is based on oases. Oases vary from small water holes, where an underground aquifer meets the surface, to the Nile valley in Egypt. These fertile areas are characterized by the many date palms.

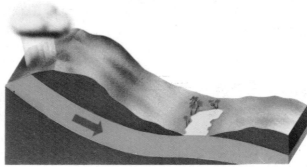

Left: Oases occur in deserts where water-bearing rock layers, or aquifers, reach the surface. Desert travellers have been known to die of thirst when water has been only a few metres away, beneath their feet.

Below: A sandstorm approaches an oasis in Algeria. Date palms are characteristic plants of Saharan oases.

Left: This photograph of a desert scene was taken at Tamanrasset, in the far south of Algeria. Desert landscapes vary and each type has an Arabic name. A bare, rocky, sandless landscape is called hammada. A sandy desert is called erg, while land covered by loose gravel and pebbles is called reg.

V **Valley** is a depression in the Earth's surface which often contains running water. Youthful river valleys are V-shaped, but glaciation often makes them U-shaped. Nearer the sea, valleys become increasingly broader and flatter in profile.
Ventifacts are pebbles shaped by WIND EROSION. At first, the pebbles have 3 faces, 2 eroded and 1 original. These are called dreikanter, a German word for '3 edges'. Eventually, only 2 eroded faces are left. These are called zweikanter, i.e. '2 edges'.

Victoria Falls, in Rhodesia, are about 120 metres high on the Zambezi River.

W **Wadi** is a valley which is only occasionally filled with water. It often occurs in desert areas and is quickly filled after rare thunderstorms.
Waterfall is a vertical fall of river water. The world's highest is SALTO ANGEL.
Watershed, or divide, is an area which divides river networks flowing in opposite directions. For example, in the western USA, the Great Divide separates rivers flowing westwards into the Pacific Ocean from others flowing eastwards to the American Plains.

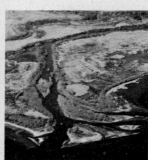

Tongariro River delta

Water table is the level below the ground above which rocks are not saturated by water. In other words, it is the top of the zone of saturation. Where the water table meets the surface, a spring may form.
Waves are ripples on the surface of the sea. They are mainly caused by the wind which heaps up the water. Along coasts, they are destructive agents of erosion.
Weathering is the process by which rocks are broken down and decomposed by the action of wind, rain and temperature changes. It is the start of DENUDATION. The two main types of weathering are mechanical and chemical.
Wind erosion is the wearing away of rock by the abrasive action of particles carried by the wind.

Y **Yardang** is a rock which has been grooved by wind erosion.

Z **Zeugen** is a resistant rock resting on undercut pillars, like a 'rock mushroom'. Wind erosion narrows the pillar, but the top resists erosion.

Finding out about the Earth is not the sole preserve of trained geologists, map-makers or expert mountaineers. A great deal can be discovered by anyone with a questioning mind and a love of the open air.

Exploration and Geology

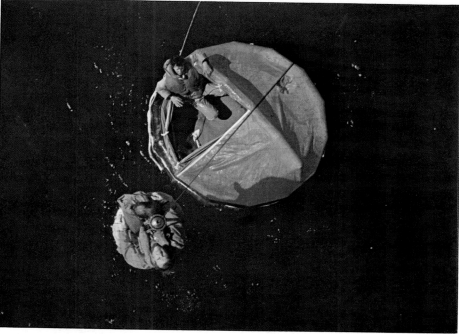

All that is needed to appreciate the geological forces around us is a keen eye and a questioning mind. Why do hills have different shapes? Why does a waterfall occur at a particular point in a river? How are pebbles rounded and polished? You can find out many answers for yourself.

Ups and downs

Walking is a leisurely way of find out about the Earth on which we live. But more strenuous exercise is involved when exploring the deepest and highest parts of the Earth. Exploring deep limestone caves attracts pot-holers (*see pages 104–105*), and mountaineering has become a popular sport in the last 100 years or so.

The 'golden age' of European mountaineering was the 1850s and 1860s, when most of the great peaks of the Alps were climbed. The period reached a climax with the conquest of the

Above left: Mountaineers, with their considerable battery of equipment, prepare to climb the Eiger, a 3,975-metre-high peak in the Bernese Alps. With modern, sophisticated equipment, climbers can scale mountains and precipitous routes which could not be attempted in the heyday of Alpine mountaineering in the mid-1800s.

Above right: A helicopter rescue team winch up the crew of a lifeboat. Helicopters are valuable in both sea and land rescue. They travel quickly and can hover over the place where they are needed even if they cannot land.

MATTERHORN in 1865. However, a number of 'firsts' remained for 20th-century climbers, especially in the great Himalayan range. The top of the world, Mount Everest, was conquered on 29 May, 1953, but there are still peaks in the Himalayas that have never been successfully scaled.

Outdoor activities have become increasingly popular in recent times and most countries have hills or mountains to climb. But pot-holing and mountaineering can be dangerous. The inexperienced should join a club and get advice on equipment, and guidance from experienced people.

In addition to the obvious need for warm and lightweight clothing and sturdy boots, mountaineers need special equipment, such as CRAMPONS (steel spikes for boots), hammers, PITONS (metal pegs to drive into rocks to secure a

Reference

A **Aerial photography** is a modern method of mapping the details of the land. It has come widely into use in the last 30 years because it is quick and especially suitable for areas where travel on the ground is difficult. Prior to an aerial survey, a TRIANGULATION network of control points is fixed on the ground. The COORDINATES and heights of these points are accurately measured. The vegetation

around the points is cut away, often in the form of a cross. Such crosses are visible on aerial photographs. In the aeroplane, photographs

Gemini 4 view of Mauritania

are taken in long strips, in such a way that each photograph overlaps the next by about 60%. The overlaps can be viewed stereoscopically – that is, an overlapping pair appears in 3-D when viewed through a stereoscope. Hence, the height of the land can be mapped, as well as land details. Very sophisticated techniques, including the use of computers, are now applied to compile maps from aerial photos.
Ammonites are extinct molluscs (boneless, mostly shelled animals) that lived in the Mesozoic era. Their

Ammonite

spiral shells are often preserved as fossils. Ammonites vary in size from less than a centimetre to over 2 metres across. Ammonites,

which resembled the modern *Nautilus*, became extinct about 63-64 million years ago.

B **Belay** is a mountaineering technique of securing oneself, or another climber, by passing a rope around a projection of rock or through a KARABINER.

C **Cartography** is the art of making charts or maps.
Clinometer is an instrument for measuring vertical angles. From the distance between 2 points and the

holding), a strong nylon rope, and sometimes a tent and sleeping bag.

To find your way up a mountain or across the countryside, you need a map. Maps are accurate representations of part or all of the Earth's surface on a flat sheet of paper. They show hills and valleys, coasts and lakes, railways, roads, bridges, towns, churches and many other features. The features are represented by conventional symbols, which are the map-makers' shorthand. The legend (key) to the map explains all the symbols. For example, railways are usually shown as black lines. Roads are normally coloured lines, but the colours and thicknesses of the lines vary according to the type of road. Rivers, lakes and seas are shown in blue, and so on. In fact, one detailed map may contain enough information to fill an entire book.

A sheet of paper the size of this page may contain a plan of a room, or a map of the world. The difference between them is a question of SCALE. For example, on a map drawn to a scale of 1:1,000,000 (one to a million), one centimetre on the map represents 1,000,000 centimetres (10 kilometres) on the ground. At this scale, southern England and South Wales could be shown on these two pages. But, to show the world, this requires a much smaller scale, such as 1:80,000,000, where 1 centimetre represents 800 kilometres. Neither of these maps can show much detail. A popular scale for topographical maps,

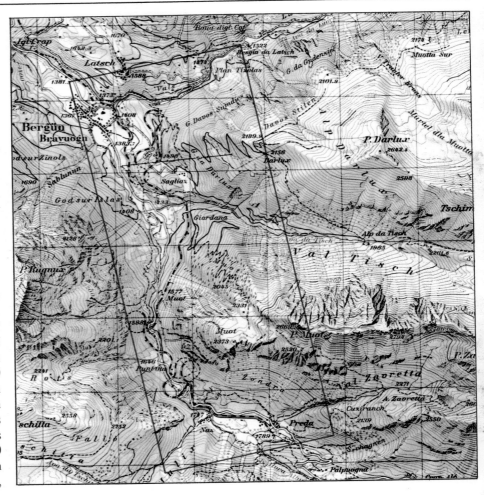

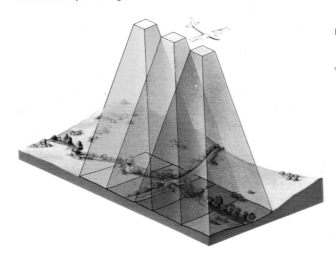

Below: Photographs taken in strips from aircraft are widely used in modern detailed map-making.

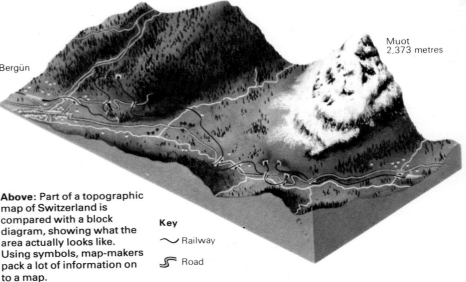

Muot
2,373 metres

Bergün

Above: Part of a topographic map of Switzerland is compared with a block diagram, showing what the area actually looks like. Using symbols, map-makers pack a lot of information on to a map.

Key

〜 Railway

ʃ Road

vertical angle between them, the difference in height can be calculated.

Compass is an instrument for finding direction. The magnetic compass, invented by the Chinese, came into use in Europe in the AD 1200s. It is based on the fact that a freely suspended piece of magnetic material will always point towards the magnetic poles. The magnetic north and south poles are situated near the true geographical poles and so small corrections have to be made to magnetic compass bearings to obtain ac-

curate directions. Mariners' compasses contain a card, to which strips of magnetic metal are stuck. The card usually floats in alcohol, which stops it swinging around. Radio compasses are rather different. They consist of a rotating antenna on an aircraft. The antenna points in the direction of radio signals transmitted from a known ground station. By measuring the direction of the antenna, the direction in which the aircraft is flying can be worked out.

Contours are lines on a map linking places of equal

height. A conical hill with even sides would be represented on a map by a series of circular contours, one inside another. River

Rock climbing equipment

valleys are represented by sharply bent contours. The contour interval (the vertical distance between contours) varies according to the scale of the map. On some maps of extremely steep mountains, HACHURES may replace contours.

Co-ordinates are lines drawn vertically and horizontally across a map. They help map-users to locate places. On world maps, the lines are lines of latitude (parallels) and lines of longitude (meridians). Lines of latitude are drawn parallel to the equator. The equator is

0° latitude, the North Pole is 90°N and the South Pole is 90°S. Lines of longitude are lines which pass through both poles. Lines of longitude are measured 180°E and 180°W of the prime meridian which passes through GREENWICH. In writing the co-ordinates of a place, it is customary to write the latitude first and the longitude second. Hence, the co-ordinates of Toronto are written: 43°39'N, 79°20'E.

Crampon is a metal framework carrying steel spikes which can be attached to climbing boots

which are used by walkers, is 1:50,000. One centimetre at this scale represents 0.5 kilometres on the ground.

Map-makers of old

The earliest known map is about 4,300 years old. It was drawn on a clay tablet by the ancient Babylonians. Modern cartography (map-making) began with the ancient Greeks. They realized that the Earth was round and they calculated its size. They were aware of the equator and the poles and they divided the world into latitude and longitude. They even made PROJECTIONS – ways of showing the curved surface of the Earth on a flat sheet of paper. The art of map-making developed in the Middle Ages and received a great impetus during the Age of Exploration (1450–1750). The first detailed geological map (of southern England) appeared in 1816.

How maps are made

Modern maps are highly accurate and this requires expensive and complex equipment for measuring distances and angles.

The mapping of an area begins with the establishment of a series of points on the ground arranged in triangles, possibly many kilometres apart. They must be intervisible through telescopic instruments, called THEODOLITES. The corner of each triangle is called a triangulation

station, often marked by a concrete pillar. To fix the exact positions of these points, surveyors must first measure the distance between two of the points. Accurate ground measurements with metal tapes take a long time. However, the positions of other points in the network can be fixed quickly by angular measurements alone, because, if you know the length of one side in a triangle and all three angles, you can then work out the length of the other two sides. This method of surveying is called TRIANGULATION.

From the network of triangulation stations, the details of the land can gradually be filled in. This is sometimes done by ground measurements but land details are increasingly being mapped by the faster method of AERIAL PHOTOGRAPHY.

There are many kinds of maps. Large-scale maps showing only a few streets are usually called plans. Small-scale maps leave out many details but they cover large areas.

Relief maps

Physical, or relief, maps show the topography (the shape of the landscape) and include hills, mountains, valleys and rivers. The heights are often shown by CONTOURS – brown lines which join places of equal height. However, other techniques may be used. For example, on small-scale atlas maps, layer tinting is a common method. This means that a colour represents all the land between two heights.

Below: The geological map and cross-section show a region called the Weald in south-eastern England. The Weald is an anticline (upfold), the top of which has been worn away. Hence, on the map, the chalk lands on the south coast are part of the same layer which outcrops in the north, where it reaches the sea in the famous White Cliffs of Dover. The chalk once extended right over the top of the anticline, but it has been removed by continuous erosion.

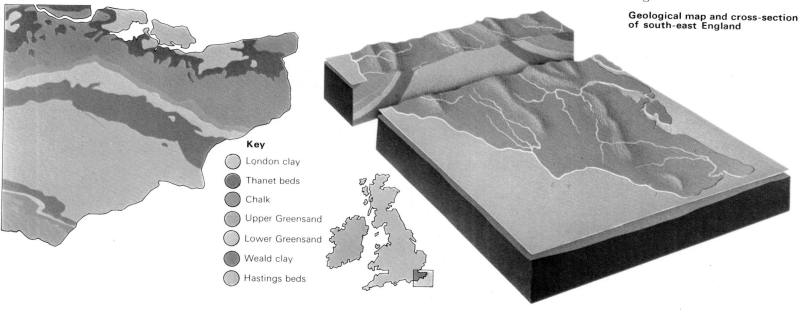

Geological map and cross-section of south-east England

Key

London clay
Thanet beds
Chalk
Upper Greensand
Lower Greensand
Weald clay
Hastings beds

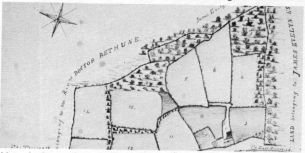

Map of a Sussex farm drawn in 1793

with leather straps. The spikes dig deeply into snow and so save time which would otherwise be spent in cutting steps.

D Drilling is boring into the ground to investigate the rock or soil types. Samples are brought to the surface through the drill as a core, shavings, or liquid for examination. Drilling is an important part of oil prospecting.

G Geodimeter is an instrument used to measure distances quickly. Geodimeters record the time taken for light waves to travel between two points. Because the speed of light is known, the distance can be worked out.

Geophysics is a study of the physical characteristics of the Earth, including magnetism, pressures, densities and elasticities of rocks. It is also concerned with the study of shock waves, generated by earthquakes or explosions. It has many applications in modern prospecting.

Geological mapping is undertaken by geologists who examine the rocks of an area and mark the outcrops on a map. They obtain information from rock outcrops and from DRILLING beneath the soil. Evidence also comes from geophysical and geochemical studies, old maps, old mines and so on. All the information available is plotted on a map and unknown boundaries between rocks of different types and ages are estimated.

Oil exploration in Bolivia

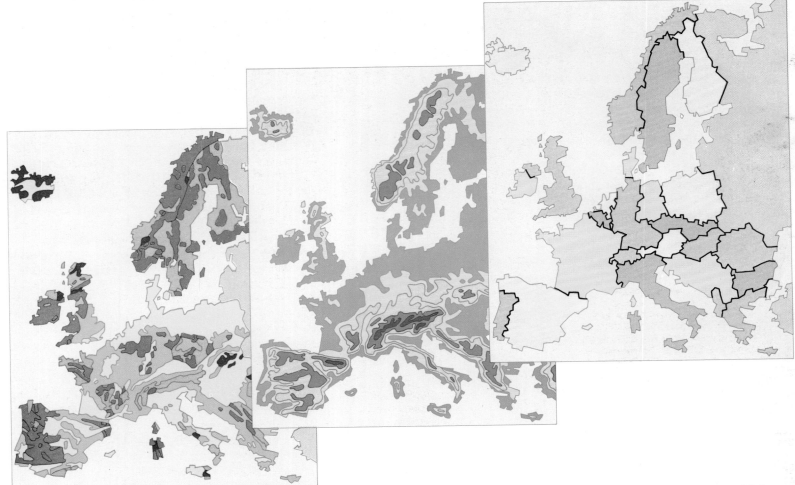

Above: Each of the 3 maps of western Europe supplies a different type of information. On the left, there is a simplified geological map, identifying rocks which outcrop at the surface. The physical map (*centre*) shows land features, such as mountains and plains. The political map (*right*) depicts the countries of Europe. A great variety of special maps can be produced, each giving information of a different kind.

Geological maps

Geological maps show, in a variety of colours, where different rocks outcrop on the Earth's surface. However, in most areas, these rocks are covered by soil. To some extent, therefore, the exact boundaries of rocks have to be estimated. Geologists, however, obtain a lot of information from DRILLING and from rocks exposed in river beds, railway cuttings, cliffs, and so on.

Many geological maps also give information about rock structures, such as the tilt of rock layers and the location of faults. They may also include cross-sections which show how the rock strata are arranged beneath the surface.

Other maps

Maps are produced for many purposes. There are road maps for motorists, charts for navigators, population maps which show the distribution of people, weather maps which are familiar from newspapers and TV, and climatic maps showing average annual rainfall and temperatures. There are also political maps with national and local boundaries, economic maps depicting the location of mines and mineral reserves, archaeological maps, agricultural maps, maps of the oceans, and even geophysical maps showing the level of magnetism or some other physical characteristic.

Europe and North America have been mapped in considerable detail and surveyors keep these maps up to date by recording changes, such as the construction of new towns and roads. But many other parts of the world are mapped in far less detail. Detailed mapping in the developing world is often undertaken only when minerals are discovered. For example, this is now happening in many countries in the Arab world.

Greenwich Observatory, in south-east London, is traditionally taken to be the place through which the prime meridian (0° longitude) passes. It is also the place from which world time zones are calculated. The time at Greenwich is called Greenwich Mean Time (GMT). Moving east or west of Greenwich, the time of day changes by one hour for every 15° of longitude, because it takes 24 hours for the Earth to rotate once on its axis (24 x 15 = 360, or the number of degrees in a complete circle). East of Greenwich, time zones are ahead of GMT, but west of Greenwich, time zones are behind GMT. When it is noon at Greenwich (0° longitude), it is midnight at longitude 180°, the International Date Line. However, the International Date Line is 12 hours ahead of GMT going east-

Greenwich Observatory, London

wards, but it is 12 hours behind GMT going westwards. Hence, at the International Date Line, there is a difference of 24 hours, or one day. The International Date Line does not follow 180° longitude exactly. Instead, it is kinked so that it avoids land. This ensures that no Pacific island has two dates on the same day!

Graptolites were small sea animals that lived between the Cambrian and Carboniferous periods. They were particularly abundant in the Ordovician and Silurian periods. Their fossils, which often look like fretsaw blades, are often found in shales.

H Hachures are lines of shading drawn on some maps to give the impression of the relief. They may be used instead of CONTOURS on maps of steep, mountain regions, where the contours would be so closely packed as to merge into one another.

I Infra-red photography allows photographs to be taken through clouds or at night. Instead of using film

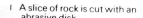

1 A slice of rock is cut with an abrasive disk

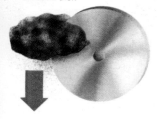

2 The slice is cemented to a glass slide with Canada balsam

3 The slice is placed on an iron plate and ground smooth using corundum powder

4 When the slice is transparent, a cover slide is then cemented over it

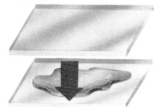

The slide is now ready for the microscope

Above: This infra-red photograph of part of San Bernardino County, California, in the USA, was taken at a height of about 15,240 metres. It shows a braided river, an Air Force base (top) and the San Bernardino mountains.

Below: This photograph of a thin slice of arkose was taken in polarized light. Arkose is a sedimentary rock, containing a high proportion of the mineral feldspar (over 25%), as well as other minerals.

Above: The diagrams show how you can prepare a thin slice of rock for study under a microscope. Even the tiniest crystals can then be identified.

Geological expeditions

Geological expeditions can be fun. In hunting for rocks and minerals, you do not require much equipment – basically a hammer, a notebook and pencil and, perhaps, a small magnifying glass. The hammer can be used to break off a piece of rock to expose a fresh surface. Notes of any details, such as where a rock was found, are very useful, because they are easily forgotten. Finally, the magnifier enables you to examine and identify small mineral grains in rocks.

Stones and pebbles on beaches are often colourful and full of geological history. The sea constantly rounds and polishes the stones, which may be so striking that they can be used in making jewellery or paperweights.

While large mineral crystals in rocks can be seen through a magnifying glass, very small crystals cannot be seen. Small crystals must be examined through a microscope. First, you must prepare an extremely thin slice of rock and mount it on a piece of glass (*see diagram*). A thin glass cap is then placed over the section and the sample is placed in the microscope. Light from below shines through the rock slice and into the lens above. The small crystals can then be identified by their shape and colour.

---

sensitive to ordinary light as in a normal camera, film sensitive to infra-red light is used.
International Date Line, see GREENWICH OBSERVATORY.

K **Karabiner** is a metal snap ring through which a rope can be passed for added security. It is often used with a PITON or rope BELAY.

L **Lapidary** is a cutter or shaper of stones and rocks, particularly to make them attractive. The stones are often put into a drum and rotated, or 'tumbled', so that they become polished.
Lens is a small magnifying glass used by geologists to examine crystals in rocks.

Rock climbing

Such lenses usually have a magnification of 10 times.
Levels are telescopic instruments used to make accurate measurements of the height of the land. Apart from the telescope, the other main feature of these instruments is a highly-sensitive spirit level.

M **Matterhorn** is a huge, rocky mountain on the Swiss-Italian border in the Alps. It is 4,478 metres above sea level and is one of the most climbed mountains in the world. It was first conquered in 1865.
Meridian is another word for a line of longitude. The prime meridian (0° longitude) passes through GREENWICH.

Matterhorn

Left: Slate is a metamorphic rock formed from shale by being subjected to great heat and pressure. It has good cleavage and is widely used in roofing.
Centre: Galena is the richest lead ore. It often has traces of silver in it.
Bottom: This specimen of calcite (calcium carbonate) is known as dog tooth spar.

Top: Rock salt is usually formed when a sea evaporates and salt layers are covered by other sedimentary rocks.
Centre: Malachite is one of the 7 main ores of copper.
Right: Haematite is an important ore of iron. This form is called kidney ore.

Identifying rocks and minerals

For beginners, a guide to rocks and minerals is a useful purchase. Such books not only describe rocks and minerals, but they also contain photographs of samples, which you can compare with those you have collected.

When identifying minerals, there are a series of tests which can be made. For example, you can find the hardness of the mineral on the Mohs' scale (*see page 72*). Colour is another useful feature, although it can be misleading if the mineral has been exposed to the air or affected by impurities. If you scratch a mineral sample, the powder, or streak, may have a different colour from the mineral itself. Each mineral also has a characteristic lustre, such as metallic, non-metallic, glassy or dull. Other means of identification include the crystal forms, cleavage and specific gravity.

Above: A selection of 6 rock specimens. The rock salt comes from the island of Elba in the Mediterranean, the malachite from Zaire in Africa, the galena from Scotland, the slate from Cornwall, the haematite from Lancashire, England, and the calcite from Mexico.

Books on minerals list each of the above characteristics for each mineral. For example, cassiterite (tin oxide) is a mineral ore of the metal tin. It has a hardness of 6-7. Its colour varies from reddish-brown to black, although yellowish samples may also be found, but the streak is white to grey. The lustre varies from adamantine (brilliant) to submetallic (a glassy, metallic lustre). The crystal form is tetragonal (pyramidal), it has imperfect cleavage and its specific gravity is exceptionally high at 6·8-7·1. Samples of cassiterite can be identified from a combination of these features. Other minerals can be identified in a similar way.

Fossil collecting

Fossil hunting in sedimentary rocks has become another popular geological hobby. For beginners, it is useful to visit your local museum and

---

Microscope is an instrument for viewing extremely small objects. Geologists often use them to view mineral crystals in rocks. A beam of light shines from below through a very thin slice, or section, of rock. The light continues up into the eyepiece or lens, through which the observer looks. The eye piece magnifies the image of the rock specimen which can then be examined in detail. It is possible to gather more information by using POLARIZED LIGHT.
Mineralogy is the study of minerals.

O **Optical properties** of a mineral are its characteristics when viewed through a microscope. The major properties are colour and how it affects (bends) light passing through it. Some minerals change colour when viewed in POLARIZED LIGHT. This property is called pleochroism.

P **Parallels** are lines of latitude. They are parallel with the equator and are measured between the equator (0° latitude) and the poles (90°N and 90°S).
Petrology is the branch of

San Francisco Bay area

geology which deals with the composition, structure and classification of rocks.

Piton is a short metal bar which can be driven into a rock fissure or ice. Used for BELAYING, they often have a snap ring attached to their blunt end, through which a rope can be taken.
Polarized light is a special light used for viewing minerals under a MICROSCOPE. Ordinary light is an energy wave that vibrates in all directions. Polarized light vibrates in one direction, because it is passed through a type of grid. Different colours show up in polarized light and this helps geologists to identify them.

Projections are ways of representing the curved Earth on a flat piece of paper. Except for maps of small areas, where the Earth's curvature is insignificant, there is no way of preserving all the Earth's features on a flat piece of paper without some distortion. For example, if you peel an orange, the skin cannot be pressed flat without tearing and distorting it. Hence, the only true map is the globe. However, many projections have been devised for particular purposes. For example, projections which

Left: Flint is a hard and fairly pure form of silica that was used to make tools in the Stone Age.
Centre: Feldspars are the most common group of rock-forming minerals.
Bottom: Shale is a common, sedimentary rock formed from layers of silt and clay, sometimes containing fossils.

Top: Quartz is one of the commonest of all minerals as it combines the 2 most common elements in the Earth's crust.
Centre: Crystalline gypsum is a softish mineral used in blackboard chalk and plaster of Paris.
Right: Granite is an igneous rock. It consists mainly of feldspar and quartz.

discover what kinds of fossil are found in your home area. The local museum or a collectors' club will also advise you on the best places to look for fossils. These places will usually be sites where rocks are exposed, such as quarries, railway cuttings, embankments, cliffs and so on. Unfortunately, searching for fossils in such places can be dangerous and great care should be taken. Do not climb up cliffs or quarry walls. It is easier and safer to look for fossils in the rocks on the ground. Always get permission to enter quarries and never disregard signs which tell you to close gates or keep clear of dangerous cliffs.

Besides a hammer, notebook and pencil and a magnifying glass, other useful items for fossil hunting are a chisel, small boxes for packing specimens so that they do not break before you get them home, a geological map to identify the age of the rocks, labels and so on.

Above: A further selection of 6 rock specimens. The quartz and granite come from Scotland, the gypsum from Cumbria, the flint from Norfolk, the feldspar from Cornwall, and the shale with mollusc laminations from the Isle of Wight.

Do not expect to find the massive bones of dinosaurs around every corner, although amateurs occasionally find spectacular and even hitherto unknown fossils. If you come across something too big to handle, it is best to notify your local museum and experts can then come along to extract it. Small fossils, however, are abundant in some areas. They include TRILO-BITES and GRAPTOLITES in Palaeozoic rocks and AMMONITES in Mesozoic rocks. Plant remains and even fossil fishes are not uncommon. For identification, you need a well-illustrated book on fossils, but if you are ever baffled, then your local museum may be able to help.

Good fossil sites are often visited by thousands of people every year. It is therefore important to avoid unnecessary damage as fossils are invaluable evidence of the past, and good sites will eventually be 'worked dry'.

---

give true directions have been made for navigators and equal area projections have been devised for geographers.

Chalk

R **Rappel** is a way of climbing down steep rocks with a doubled rope. The rope is passed around a projection at the top of the rock, taken beneath the climber's legs, over his shoulder and across his back. Friction on his clothes helps to regulate the speed of his descent.

S **Scale** is the representation of long distances on the ground by short distances on a map. Scale may be stated as a representative fraction, such as 1:10,000. This means that 1 cm on the map equals 10,000 cm (100 metres) on the ground. Many maps have graphic scales, such as a line divided into kilometres.

T **Tellurometer** is an instrument used to measure distances quickly. It records the time taken for electromagnetic waves to travel between 2 points. It can be used even when the visibility is poor.
Theodolites are telescopic instruments used to measure horizontal and vertical angles in land surveying. *See also* LEVELS.

Triangulation is the first stage in mapping an area. It involves fixing (accurately) the positions of a network of points arranged in triangles. After the measurement on the ground of the distance between 2 of the points (called the base line), the rest of the measurements are angular.
Trilateration is a method of surveying an area by measuring distances, rather than angles as in TRIANGULA-TION. It is based on the mathematical principle that if you know the lengths of 3 sides of a triangle, then you

can work out the angles. Trilateration has been used since the invention of instruments, such as GEODIMETERS and TELLUROMETERS.
Trilobites are a group of extinct sea animals which lived between the Cambrian and Permian periods.
Tropic is one of the 2 corresponding parallels of latitude on the globe, representing the northernmost and southernmost places where the Sun is overhead once a year. The Tropic of Cancer is situated at 23½°N while the Tropic of Capricorn is 23½°S.

Ours is a complex world, confronting man with colossal problems. However, with continued research and constant vigilance, the Earth can remain a wonderful habitat for the human race. Lack of understanding could lead to global disaster.

Man and the Earth

Life was once simpler than it is today, although it was also harder and life expectancy was shorter. Early man lived by gathering food, hunting and fishing. About 10,000 years ago, AGRICULTURE was invented in what is now the Middle East. Crops were grown, animals reared and permanent homes built. The landscape began to change, although at first it was hardly noticeable, because the world's population was small. It has been estimated that, in about 8000 BC, there were only about eight million people and, by about AD 100, there were still only about 300 million.

As the world's population increased, the countryside was changed by farming, mining and the building of larger settlements. During the late 1700s, the INDUSTRIAL REVOLUTION began in Britain, which soon became the 'workshop of the world'. Machines replaced hand labour and coal and steam replaced animal power.

Coal was used to smelt iron, which enabled steam-powered machines to be built. Factories

Above left: This forest clearing in Sri Lanka has been stripped of its natural vegetation to provide farmland. When the soil is exhausted, the people will abandon this plot, clear a new area and move their homes. This wasteful method of farming, called shifting cultivation or 'slash and burn' is widely practised in many tropical areas.

Above right: The icy continent of Antarctica has no permanent population. The only visitors are explorers, scientists, fishermen and whalers. About 90% of the world's land is unsuitable for crop growing, although few places are quite as bleak as Antarctica.

increased in number and cities grew in size. Industrialization spread to countries such as Belgium, France, Germany and the USA by the mid-1800s. In the late 1800s, Canada, Japan, Russia and Sweden also became industrial nations.

The world's population continued to grow at an even faster rate, from 900 million in 1800, to 1,650 million in 1900, 2,000 million in the 1920s, 3,000 million by 1960 and the 4,000 million mark was passed in 1977.

Farming

When man put down his bow and arrow to take up the plough, he started the agricultural revolution. The plough was invented in about 9000 BC. The original wooden stick was later improved when the blades were made of metal and the plough was adapted so that it could be pulled by draught animals. IRRIGATION was developed, especially in the Middle East, where

Reference

A **Agriculture** is the farming of the land to produce crops. Usually, it also includes animal husbandry. Agriculture is the world's leading industry, employing more than half of the world's people. But in countries where farming is mechanized, comparatively few people work on the land. For example, only 6% of North Americans work on farms as compared with 74% in Africa.

B **Base metals** include copper, lead and zinc. They are used in manufacturing industries.

Ploughing, sowing and harrowing

Breeding new varieties of plants and animals is an important part of scientific agriculture. Plant breeders produce mutations (variants) which may result in bigger, stronger and more productive varieties. The modern pig is an example of animal breeding. Its ancestor, the wild boar, was heaviest around its shoulders. But the best meat comes from the back end. Hence, only pigs with relatively larger hindquarters and smaller shoulders were used in breeding.

Bromine is an element used in the photographic and pharmaceutical industries. It is largely obtained from seawater.

C **Cities** are large concentrations of population, usually associated with manufacturing, commerce and trade. In developed nations, most people now live in cities and towns, but 60–90% of the people in DEVELOPING NATIONS live in rural areas. Cities provide many amenities, but their inhabitants also face such problems as noise, air pollution, traffic jams and, sometimes, high levels of crime. The 10 largest cities are as follows:

Tokyo (Japan)   11,702,000
Shanghai (China) 10,000,000

Ancient ploughs

Seed drill

First motor-driven tractor

Below: Rice fields near Mysore in southern India. There are many varieties of rice. The commonest one in rainy or irrigated areas is called lowland, swamp or wet rice, because it is planted in flooded fields. About half of the world's people depend on rice as their basic food. China is the leading producer.

it played an important part in the growth of early civilizations. However, in many places, farming was primitive. People cleared a patch of land, usually by burning the vegetation. They then grew crops until the patch was exhausted and moved on to another area. This 'slash and burn' or 'shifting' cultivation is still practised in some DEVELOPING NATIONS.

CROP ROTATION, which helps to preserve the fertility of the soil, was introduced into Europe in the Middle Ages, whereas military conquest and exploration led to the introduction of new crops from Asia and the Americas.

Farm machinery

Farm machines were first used extensively on the great flat wheatfields of the USA and central Canada. These areas were called the 'bread basket of the world', because they produced so much of the world's wheat. Inventions, such as the tractor and the combine harvester, replaced

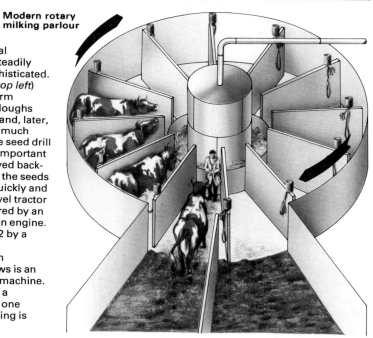

Modern rotary milking parlour

Above: Agricultural techniques have steadily become more sophisticated. Ancient ploughs (*top left*) were pushed by farm labourers. Metal ploughs drawn by animals and, later, by machines were much more effective. The seed drill (*top right*) was an important step forward. It saved back-breaking work and the seeds were distributed quickly and evenly. The early Ivel tractor (*centre*) was powered by an internal combustion engine. It was made in 1902 by a British engineer.

Right: This modern roundabout for cows is an automatic milking machine. The cows stand on a turntable and after one revolution the milking is over.

Mexico City (Mex.) 8,942,000
Buenos Aires
  (Argentina)   8,775,000
Seoul (Korea)   8,684,000
Cairo (Egypt)   8,143,000
Peking (China)   8,000,000
New York (USA)   7,896,000
Moscow (USSR)   7,734,000
London (UK)   7,379,000
Conservation is the saving of raw materials, energy and land for the future. For example, this can be done by reducing consumption or by using what we have more efficiently.
Copper is a reddish-brown metal used in industry (pipes and electrical wire).

Crop rotation is a system used to avoid exhausting the soil when crops are grown. There are several methods. For example, a farmer with

3 fields might leave one fallow (empty) each year; grow wheat or rye in a second; and the third might be sown with barley, oats or

peas. Every year, the crops would 'rotate', so that no one field grew the same crop for 2 years in succession. This system was introduced into Europe in the Middle Ages.

D Demography is the study of population.
Desalination is the making of fresh water from sea-water. This may be done by distilling seawater by heating it; freezing the water into ice; or by using electrical processes. All the methods are costly and are economic only in desert regions which

are desperately short of water.
Developing nations are countries which are poorer than developed nations.

Cultivation goes right up to the cliff edge

Star of Africa diamond

Above: Open-cast mining for iron ore began at Iron Knob in South Australia in 1900. Australia is now the world's third largest producer of iron ore.

Right: The diagram shows a plan of a modern coal mine. World coal production has remained static in recent years, because oil and natural gas are cheaper.

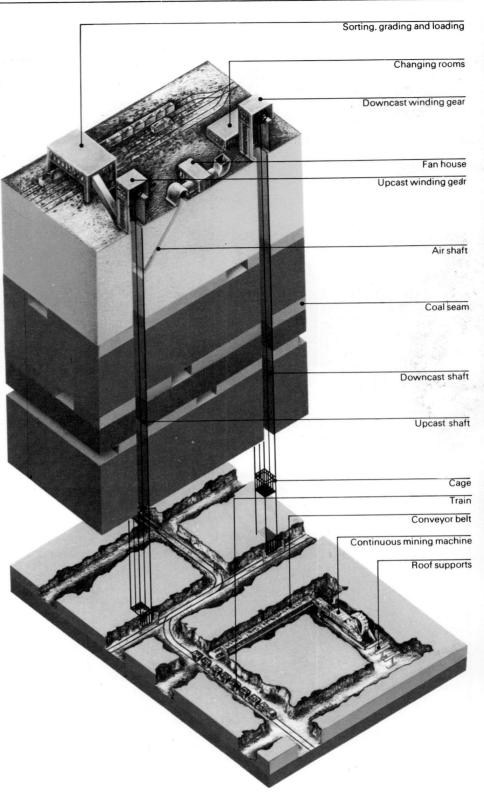

Sorting, grading and loading

Changing rooms

Downcast winding gear

Fan house

Upcast winding gear

Air shaft

Coal seam

Downcast shaft

Upcast shaft

Cage

Train

Conveyor belt

Continuous mining machine

Roof supports

the horse and scythe on many farms. Today, farming in developed nations is highly mechanized.

Scientific farming

Scientists have done much to help farmers to raise food production. FERTILIZERS which replenish the nutrients in the soil are now widely used. Diseases and pests can be controlled by spraying crops with FUNGICIDES and other PESTICIDES. More crops are grown because of new farming techniques and more productive animals have been bred. In the last 30 years or so, there has been a 'Green Revolution' in many areas. This latest agricultural revolution is based on the scientific BREEDING of new varieties of food crops, which are better adapted to particular conditions and which give higher yields. For example, a new variety of maize was introduced into Mexico in 1940. Maize was then imported but, by 1960, Mexico had become a maize exporter.

Generally, in developing nations, a high proportion of the people work on the land and manufacturing plays only a small part in the economy.

Diamond is the hardest known natural substance. It has a value of 10 on the Moh's scale of hardness (*see page 72*). It is used as a gem in jewellery and it is an important industrial abrasive, often used in drills.

Dredge is a machine used in mining to collect loose material from a lake bottom or sea bed by sucking or scraping it up into a bucket.

E Ecology is the branch of biology dealing with how and where living organisms live and how they interreact with, and are dependent on, each other. Recently, the term has been applied to the study of how pollution and other factors affects the habitat of living organisms.

Ecosystem is a group of organisms living in a particular ENVIRONMENT. If the environment is changed, the ecosystem is upset.

Effluent is liquid waste material, such as sewage or industrial waste, commonly seen as foam on rivers.

Environment is a term which refers to the conditions in a particular place in which plants, animals and people live.

Irrigated land in southern Israel

F Fertilizers contain the chemicals needed by the soil if crops are to be grown. The main ones are phosphates, containing phosphorus; potash, containing potassium; and nitrates, containing NITROGEN. Other chemicals include lime, magnesium and manganese. Some fertilizers consist of mixed chemicals.

Fish is an excellent, protein-rich food. The most common varieties in northern seas include cod, haddock, halibut, herring, mackerel and whiting. Even more varieties, such as anchovy, gurnet, red snapper and tuna, are known in southern seas. Freshwater lakes and rivers provide chub, perch, salmon, trout and eels.

Food can now be stored much longer. It once had to be salted and eaten quickly, but it can now be frozen, dried or canned. Such crops as soya beans have been found to be cheap and healthy and they can be used instead of meat. However, despite all the improved methods of farming, many people still do not have a balanced diet and famines occur. Natural disasters, such as cyclones and droughts, still lay waste large areas in Asia and Africa.

Minerals and fuels

The 20th century has given us many luxuries, such as electricity, cars, aircraft and man-made fibres, such as nylon. To produce these luxuries, minerals and energy are consumed in enormous quantities.

Natural supplies of fossil fuels take millions of years to accumulate. FOSSIL FUELS include coal, which accounts for 29 per cent of the world's energy supplies; oil, 45 per cent; and natural gas, 19 per cent. Coal was formed by the burial of great forests which grew mainly in the Carboniferous period, about 300 million years ago. As more and more rock strata formed on top of the rotting vegetation, so the quality of the coal increased. From peat, it became brown coal (lignite), then bituminous coal and finally anthracite, the highest quality coal.

Below left: An oil rig in the Fulmar oilfield of the North Sea. Part of the legs of the semi-submersible drilling platform, named *Ocean Voyager*, are below water level, but essentially the platform floats like a ship. It is due to be producing 150,000 barrels per day by 1981.

Below: The diagram shows ways in which oil and gas become trapped in porous rock layers enclosed by impervious rocks. The commonest structure is the upfold, or anticline. Other structures include the tops of salt domes and fault traps.

Petroleum (crude oil) and natural gas were formed by the accumulation and decay of millions of tiny marine plants and animals. This process produces the hydrocarbons (hydrogen and carbon), which make up oil and gas. The fluid materials are trapped in porous rocks, called reservoirs, often in arch-like anticlines.

One of the world's major problems is concerned with fossil fuels. Although considerable coal reserves exist, it has been estimated that the known oil resources would be used up in less than 30 years at the present rate of consumption. This prediction does not take account of any new reserves that may be discovered. However, OIL SHALE, a rock which yields oil when crushed, is now being considered as a source of fuel although its extraction is expensive.

Before it can be used, crude oil has to be processed in a refinery. In addition to petrol for cars and paraffin, the refineries make a great variety of other by-products, including plastics, chemicals, fertilizers, soaps and detergents, and even perfume! Coal is similarly versatile, yielding many by-products.

There are other sources of energy. Windmills and water-wheels, although old-fashioned, are still used in some parts of the world. HYDRO-ELECTRIC POWER, generated by fast-flowing streams or by water rushing over a waterfall,

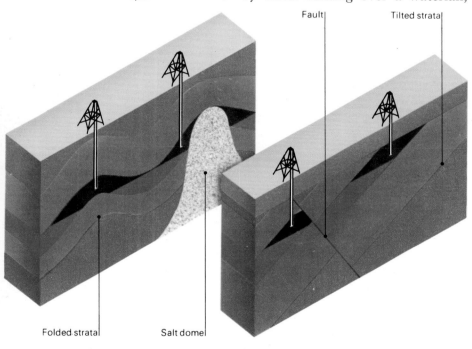

Genissiat power station on the Rhône

Fossil fuels include coal, petroleum (crude oil) and natural gas, formed by the decomposition and burial of living matter.

Fungicides are chemicals which kill the fungi that attack plants. Harmful fungi have many names, including blight, mildews, moulds and rusts. Fungicides include Bordeaux mixture (copper sulphate and lime) and mercury compounds, such as calomel (mercuric oxide).

Geothermal power is energy generated by heat within the Earth. For example, in New Zealand, electricity is generated by the natural steam which rises from cracks in the ground.
Green belts are areas around towns and cities where building is not allowed. They form pleasant recreation areas.

Hydro-electric power is largely generated by fast-flowing water in rivers and waterfalls. The rushing water is diverted so that it turns the blades of a turbine, which generates electricity. Many rivers have been dammed to create a head of water sufficient to drive a turbine. See also TIDAL POWER.

Industrial minerals are non-metallic minerals, such as limestone, potash and quartz.
Industrial nations are those countries with well-established manufacturing industries, such as iron and steel, car production, chemical works, and so on.
Industrial Revolution was a period in history when certain countries developed machines and factories. It began in Britain in the late 1700s and spread to other countries in the 1800s. Some DEVELOPING NATIONS, which have traditionally depended on farming, are now undergoing industrial revolutions.
Iron is one of the basic raw materials in the modern world. It is mined as iron ore and is often processed into either cast iron or steel.

accounts for about six per cent of the world's energy. It is particularly important in mountainous countries, such as Norway, Switzerland and parts of Africa and the Americas.

Nuclear energy

The most recently-developed form of energy is nuclear, or atomic, power. This uses the energy generated by a nuclear reaction. Although it is efficient, objections have been raised to the development of NUCLEAR POWER STATIONS, because of the danger of radioactive contamination.

Industrialized nations

Consumption of energy is a guide to the degree of industrialization in any area. The highest energy-consuming areas are North America, the USSR, Australasia and Western Europe. The rest of the world consumes much less energy, although some parts of the other continents, notably Japan, are extremely high consumers. Steel production is another indicator of industrialization, because it is essential in many manufacturing processes. The top steel-producing nations are the USA, the USSR, Japan, West Germany, Britain, France, China and Italy.

Food from the sea

FISH is a valuable food because it contains PROTEINS, valuable fats, minerals and vitamins.

The great fishing grounds of the world occur where PLANKTON (microscopic plants and animals that many fish feed on) is abundant. The world's leading fishing areas are the shallow waters of the continental shelves of the Atlantic Ocean and the Bering Sea in the North Pacific Ocean. About 66 million tonnes of fish are caught each year, with Japan, the USSR and China sharing about 40 per cent of the total.

In recent years, fishing has become more sophisticated. Equipped with radar, radio, computers and sonar (echo-sounders used to locate shoals of fish), trawlers can catch hundreds of tonnes of fish a day. Some large 'factory' ships even have facilities on board to process and freeze the fish on the same day they are caught.

Unfortunately, like everything else on Earth, the numbers of fish are limited. Some areas have been fished so much that fish have become scarce. Fish-rich areas, such as the seas around

Above: The great steel-making region of Völkingen, in the Saarland district of West Germany, used local supplies of coal and iron ore to establish the industry. Local coal is still used, but the iron ore is now imported.

Below: A car is made up of a great variety of materials, including many kinds of metals and alloys, chemicals, plastics and rubber. Each material has special properties which make it suitable.

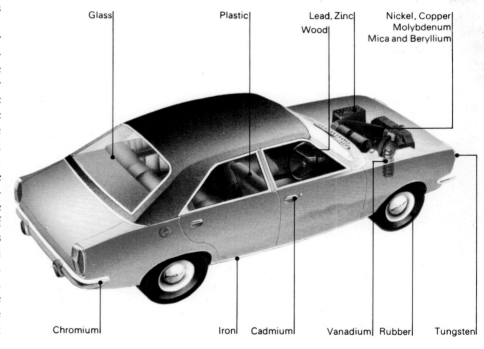

Glass   Plastic   Lead, Zinc   Wood   Nickel, Copper Molybdenum Mica and Beryllium

Chromium   Iron   Cadmium   Vanadium   Rubber   Tungsten

These, in turn, are manufactured into everything from a bridge to a tent pole.
Irrigation is the watering of the land by bringing in water via ditches, pipelines or water sprinklers. Through irrigation, many otherwise dry areas can support a wide range of crops.

An early ironworks in Shropshire, England

L **Lumbering,** or forestry, is an important industry. Timber has many uses, not only for construction, furniture and paper, but also for many varied products made by chemical industries, including plastics, paints, medicines, perfumes and explosives.

M **Magnesium** is a metal used in manufacturing lightweight alloys. In the 1930s, Germany was the main producer and so American and British scientists found a way of extracting it from seawater. Today,

over 60% of the magnesium comes from the sea.
Malnutrition is caused by under-nourishment, through lack of food or an unbalanced diet. Many people in DEVELOPING NATIONS suffer from malnutrition.
Mineral processing is the treatment of ore. It often involves the crushing and sorting of the ore, followed by the separation of the valuable metal or mineral parts. The waste is called gangue.
Mining is the industry in which valuable minerals and metals are extracted from

the Earth. The 2 main types of mining are: underground (digging shafts and tunnels into the ground); and open-cast (surface mining).

N **Natural gas,** see FOSSIL FUELS.
Nitrogen makes up 78.09% of the air. Plants need it to survive and it is distributed in the nitrogen cycle, whereby nitrogen in the soil is taken in by plants; the plants are eaten by animals and when the plants or animals die, the nitrogen is returned to the soil (*see page 98*).

Iceland, have been fished by so many nations that the Icelanders, whose economy depends largely on fishing, have complained. Differences have arisen, especially with Britain, about the right to fish in Icelandic waters and how much any nation can take. These disagreements led to conflicts, called the 'Cod wars'. Today, most nations are trying to regulate their catches in an attempt to preserve what is called 'fish stocks'. For example, the holes in nets have to be large enough to allow small, young fish to escape and so continue to breed.

One fairly recent development is fish farming. Millions of tonnes of fish are now kept in enclosed ponds or protected lakes. There, they can be bred and reared under controlled conditions. Another factor is individual taste,

Above left: Three Norwegian boats are seen here fishing in the North Sea.
Above right: Fish farming at Lok Ma Chau in Hong Kong near the border with China, is important here because fish is a popular food in the region.
Below: The diagram shows 3 types of nets used by fishermen: the enclosed otter trawl net (*left*): the purse seine net, which is drawn around the fish (*centre*); and the long gill net (*right*). The size of the holes in the nets are usually regulated, so that small fish can escape and continue to breed.

because it may be necessary, as popular species become expensive and scarce, to introduce other species on to the market.

Energy from the sea
The sea is a great natural storehouse of energy. For example, TIDAL POWER is already used, at the estuary of the River Rance in France, to produce hydro-electricity and this method may well be applied in other areas.

Minerals from the sea
A great wealth of valuable mineral resources exists in ocean water, and on or below the ocean floor. Almost every known element is dissolved in seawater. It has been estimated that there are more than 10,000 million tonnes of gold dissolved

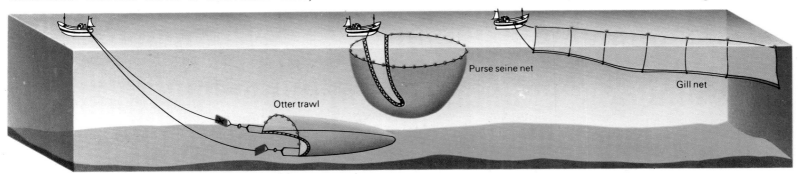

Otter trawl   Purse seine net   Gill net

Nuclear power stations produce energy from nuclear reactions. Radioactive material, such as URANIUM, breaks down, giving off radioactivity and energy. This energy is used to produce electricity. Great care has to be taken to keep the dangerous materials at the stations under control.

O **Oil shales** are rocks which, when crushed, yield oil and gas. Oil shale mining is limited, at present, to northern Canada.
Organic farming is a method of growing crops without using chemical fertilizers or sprays. The soil is enriched with natural organic material, such as animal manure.

Nuclear power station on Anglesey, Wales

Oxygen is the gas that plants and animals, including man, breathe in and that plants give off when making food. It makes up 20.95% of the air. Many polluted rivers and lakes cannot support life, because the waste products take up all of the oxygen.

P **Pesticides** are chemicals that are applied to crops and to the soil to destroy any harmful pests, such as insects, weeds, fungi, rabbits, mites, roundworms and so on. Some pesticides are sprayed onto the land, often from an aircraft, while others are spread by hand. A few, such as DDT, have been proved to be harmful to man and live-

stock and so they have been withdrawn from the market.

Smog over Los Angeles

Above: Many manganese nodules are found on the sea bed. They also contain cobalt, iron, nickel and copper. The scale (*bottom*) is in millimetres.

Right: Salt, left behind after water from salt lakes has evaporated, is here being collected near Dimboola, in Victoria, Australia.

in seawater. But it is far too expensive to extract and, at present, only salt, MAGNESIUM and BROMINE are produced from seawater.

Another source of minerals on the ocean floor takes the form of so-called manganese nodules. These potato-shaped nodules, which also contain cobalt, copper, iron and nickel, are abundant in large areas, particularly in the Pacific Ocean. Several countries are planning to mine them in the near future.

Nearer land, vast quantities of valuable minerals are found on the sea bed. They include DIAMONDS, gravel, sand, tin and titanium, which are scraped or sucked up by dredges. Beneath the shallow waters of some continental shelves are reservoirs of oil and natural gas. However, the cost of exploitation is high.

The sea's other resource is water. Removing salt from seawater, DESALINATION, is a costly process, but it is worthwhile in desert countries which have little rainfall. For example, the oil-rich nations of Kuwait and Saudi Arabia have invested in large desalination plants.

Pollution and the future

In recent years, there has been much discussion about the effects of mining, industry and population growth on the Earth's surface. Ever since man started to clear forests to create

Below: To mine manganese nodules, engineers plan to use dredgers which will operate like giant vacuum cleaners, sucking up the nodules.

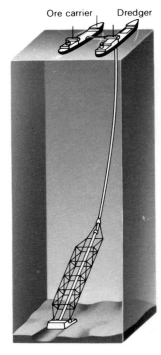

Ore carrier   Dredger

farmland, he has been upsetting nature. Now that the world's population is so large, nature is unable to cope with the changes. The ECOSYSTEM has been upset.

Pollution includes many things, such as the disfigurement of the landscape by industries or abandoned cars, noise from radios and traffic, the poisoning of seas and rivers by industrial wastes and sewage, air pollution, and so on.

Some cities, such as Los Angeles in the USA and Tokyo in Japan, have a reputation for dangerous SMOGS. Smogs occur when the top layer of air over a city is warmer than the lower layers. This causes a TEMPERATURE INVERSION. The lower layers of air cannot move upwards, so the polluted air, containing dirt particles and gases, such as those released by car exhaust pipes – carbon monoxide, nitrogen oxide and hydrocarbons – is trapped. The sunlight even breaks down some harmless gases into harmful ones, forming irritating PHOTOCHEMICAL FOG.

At one time, industry flushed its untreated wastes directly into rivers and cities did the same with untreated sewage. This killed off all life in some rivers and Cuyahoga River, in Ohio, USA, actually caught fire once, because it contained so much oil! The same is true of lakes and seas, especially those with extremely small tides. For example, parts of the Mediterranean Sea are now

---

Petroleum, see FOSSIL FUELS.
Photochemical fog is caused when bright sunlight converts harmless pollutants in the air into harmful, irritating compounds. Photochemical fog occurs in sunny cities where air pollution is severe. It makes people's eyes water.
Plankton is a term for small animals (zooplankton) and plants (phytoplankton) in the sea, on which most fish and other sea animals feed. Plankton is particularly abundant where water layers mix. Such areas are good fishing grounds.

Plastics are man-made materials that are made into a great variety of objects. They have been used as substitutes for many other materials, such as metals, glass and wood.
Population. The 10 most populated nations, with 1978 estimates of their populations are as follows:

China	857,794,000
India	636,574,000
USSR	261,000,000
USA	218,778,000
Indonesia	130,000,000
Brazil	117,080,000
Japan	114,996,000
Bangladesh	82,479,000
Pakistan	76,775,000
Nigeria	68,600,000

Population explosion is a recent term used to describe rapid world population growth. The world's population doubled between 1925 and 1975 but, if present trends continue, it will double again in only 37 years. The average rate of increase is now about 1.9% per year. However, in Africa and Latin America, the rate is even higher at 2.7%. Europe, not including the USSR, has the lowest rate of increase – 0.6% per year. At these rates, the populations

of Africa and Latin America will double in only 26 years, but it will take 115 years for this to happen in Europe. The population explosion has led to fears of over-crowding and many nations are trying to control their birth rate by massive health and education programmes.

Abandoned tin mine in Cornwall, England

Left: Air is polluted by industrial smoke, dirt and gases, as well as other things, such as the exhaust fumes from cars. Scientists have recognized that smog is a killer and many governments have begun to control air pollution. For example, in some areas, only smokeless fuels may be used.

Left: The landscape is disfigured by rubbish dumps, like this dumping ground for old cars. Land pollution is not only a problem for those who love the countryside, but such dumps also represent a waste of raw materials. The materials in these cars could be recycled.

Left: Rivers have been poisoned and made lifeless by household and industrial wastes and sewage. This pollution extends to coastal waters, around estuaries and industrial ports. By damaging the environment, we are upsetting the delicate balance of nature, which we do at our peril.

lifeless and unhealthy to swim in. Great oil tankers may cause problems. In 1978, the tanker *Amoco Cadiz* went aground off the coast of northern France and spilled about 250,000 tonnes of oil into the sea. The oil drifted over the surface of the sea as an 'oil slick'. It did much harm to fish and sea birds, besides disfiguring unspoilt coastlines. Chemicals and detergents were sprayed onto the slick to break it up.

Well-publicized examples of pollution capture the public's imagination, but less obvious examples can also do great damage. For example, the construction of the Welland Canal in the early 1900s, connecting Lake Ontario with the upper Great Lakes, had a startling effect on the fish population. Because of the canal, parasitic lamprey were able to bypass Niagara Falls and enter the upper Great Lakes for the first time. The lampreys devoured the fishes in the lakes. Catches fell drastically and swimmers, too, were attacked, demonstrating just how delicate the balance of nature is.

Cleaning things up

For generations, man has been dumping his waste without thought of polluting the environment. Only now are attempts being made to control pollution. For example, most countries now have laws to prevent air pollution. In the early 1950s, London was notorious for its thick fogs, which claimed thousands of lives every year. In 1956, however, the Clean Air Act restricted the burning of anything other than smokeless fuels. Today, London is one of the cleanest industrial cities in Europe. In 1968, the United States' government made a law to ensure that devices to reduce the amount of pollutants in car exhaust gases were fitted on cars. For developing countries on the brink of industrialization, the problems of industrial pollution and the costly remedies still lie ahead.

Similar laws check other kinds of pollution. For example, some countries ban the sounding of car horns after a certain time and do not permit mine-working at night. Industrial buildings have to be blended into the countryside and GREEN BELTS have been established around many cities. Many areas even have a pollution index and pollution levels are checked as regularly as temperatures. Fines are levied to ensure that people obey the laws.

Prospecting for minerals was once largely hit-and-miss. Today, however, many scientific methods are used especially geochemical and geophysical studies. Other information comes from space satellites.

Proteins are an essential part of all animal and vegetable life. They are found particularly in beans, butter, cheese, eggs, fish, meat and milk.

R Reafforestation is the replanting of trees which have been cut down. It can control SOIL EROSION.

Recycling is the process whereby raw materials such as glass, paper and metals may be reclaimed and re-

Prospecting for oil in Alaska

used, once the original product is no longer useful.

Resources are valuable materials available in any

area. Reserves, on the other hand, are available materials that can be mined or extracted easily and at a profit.

S Smog is the result of an area of the atmosphere becoming heavily polluted by small dirt particles and poisonous or unpleasant gases. It is largely caused by discharges from factories and fumes from car exhaust pipes. It is particularly severe when TEMPERATURE INVERSIONS occur. When this happens, the top layer of air over a city becomes heated and acts like a giant blanket,

preventing the lower layers from rising. The fumes from the city are then trapped and bright sunlight even converts some of the harmless pollutant gases into PHOTOCHEMICAL FOG.

Soil erosion is a consequence of man's misuse of the land, by bad farming. Overfarming robs the soil of its fertility and makes it less stable, and overgrazing, destroys the protective plant cover and leaves the soil open to the elements. Soil erosion occurs quickly and should not be confused with the slow processes of natur-

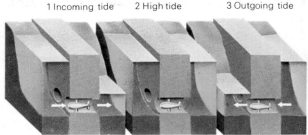

1 Incoming tide   2 High tide   3 Outgoing tide

Tidal power

Water power

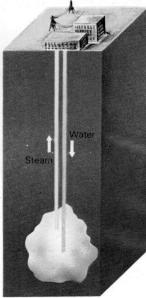

Water

Steam

Geothermal power

Wind power

Resources and the future

Much care and good sense will be needed to conserve the Earth's resources, but already various proposals have been made to prevent world shortages of materials and energy.

For instance, minerals and metals may be used more than once by RECYCLING them. Copper used in the making of a car does not disappear. It has simply been borrowed until the car is scrapped. The copper and other metals can be separated and re-used. Even rubbish dumps have their uses. In the USA, about 250 million tonnes of urban waste is collected each year. This waste could yield 12 million tonnes of iron, 1 million tonnes of such non-ferrous metals as copper, aluminium, zinc and lead, and 15 million tonnes of glass! Substitution is also important. For example, when copper becomes scarce and expensive, aluminium may be substituted. We have seen recently how PLASTICS can replace materials such as paper, glass and wood.

Nobody can say when minerals, metals and energy will run out, partly because of substitution and recycling, and partly because we cannot forsee future technology. For example, new improved PROSPECTING methods have uncovered great mineral treasures. In fact, the known reserves of many metals and minerals are now greater than they were 20 years ago. However, the world's supplies of oil, natural gas and coal are certainly disappearing rapidly. Other power sources must clearly be developed – SOLAR, WAVE, TIDAL, HYDRO-ELECTRIC and GEOTHERMAL POWER. Nuclear power is also being developed.

Other planets may, one day, yield resources but, even without them, the wealth in our seas has been barely touched. To ensure man's future on Earth, we must plan to consume less, practise conservation and continue to invest in scientific research. Recognizing a problem is the first step to solving it.

Above: The diagram depicts sources of power which may replace fossil fuels. Tidal energy is already being exploited in France. Hydro-electric power is the main source of energy in many mountainous countries. Geothermal power is obtained by using the Earth's heat to turn water into steam. Wind power also may be further developed.

Below: Solar panels collect the energy of the Sun on a mountainside in the Pyrenees. Solar energy might be developed in cloudless deserts.

al, geological erosion. The main agents of soil erosion are rain, surface run-off and the wind. To combat soil erosion farmers use many techniques, including protecting the soil's fertility with FERTILIZERS and the practise of CROP ROTATION, ploughing along land contours, and building terraces (flat steps) down hillsides. On steep slopes, REAFFORESTATION can control soil erosion.
Solar energy is the heat energy of the Sun. It can be collected and converted into other forms of energy. For example, solar cells on

spacecraft convert solar energy into electricity.
Staple foods are basic food stuffs. They include grains, such as barley, maize, rice, rye and wheat, types of peas and beans (soya bean being a meat sustitute as well), and tubers (potatoes).
Subsistence farming is the farming of crops or the rearing of livestock for the consumption of the farmer and his family. There is usually no surplus for sale and subsistence farmers have a low standard of living. It is still common in DEVELOPING NATIONS.

T Temperature inversions occur when the normal temperature decrease with height is reversed and the air close to the ground becomes cooler than the air higher up. This may occur when a warm air mass moves over cold land and the lower air layers are chilled. The cool, lower air cannot rise and, if it is polluted, it remains for some time as SMOG. The 1952 London smog, caused by a temperature inversion, killed 4,000 people.
Third World is a term used for developing countries

which do not belong to the communist bloc.
Tidal power is generated by the rise and fall of tides. Where tides are high, such as in estuaries, the movement of the water can drive turbines which generate electricity. The world's first tidal power station was opened in 1966 on the River Rance estuary in Brittany, France.

U Uranium is a radioactive element often used to produce NUCLEAR POWER and atomic bombs. The main sources are in Canada,

the USA, Australia and South Africa.
Urbanization is the trend towards living in cities.

Malnutrition in Bangladesh

Acknowledgements

Contributing artists
Alastair Campbell, Robin Gibbons, David Hardy, Elaine Keenan,
Abdul Aziz Khan, Edward Kinsey, David Mallott, Nigel Osborne

The publishers also wish to thank the following:
Aerofilms 109TL
Heather Angel 105T
Australian News and Information Service 17BR
Barnaby's Picture Library 72BL, 73R, 75, 76B, 77BR, 80, 81BL, 85B, 86B, 89B, 93BL, 93BR, 95R, 99R,
100B, 103BR, 104, 107BR, 108B, 114, 115R, 117BL, 117BR, 121BR, 125BL
John Baxter 63T
John Bethell 17C
Bodleian Library 67T, 69R
Paul Brierley 117C, 118-119
British Antarctic Survey 77T
British Broadcasting Coropration 28B, 46BL
Bruce Coleman Ltd: Norman Tomalin 50T: Chris Bonington 79TL
Colorific 88T
Crown Agents 49T
Daily Telegraph Colour Library 128T: Alex Lowe 113TL
C.M. Dixon 70T
Doubleday & Co. Inc., from the Cosmic Connection © 1973 by Carl Sagan 44R
Robert Estall 6C, 101R, 109B
Mary Evans Picture Library 74R
Fotomas Index 19B, 20L, 47B, 70B: British Library 6T
Feature-Pix 125TR
Geoslides 120TL
Hale Observatories: California Institute of Technology and Carnegie Institute, Washington 7B, 11T,
19, 31T, 43T, 46T
Robert Harding Associates 76T: Sybil Sassoon 85T: Christina Gascoigne 88C, 121T: Wally Herbert
120TR: George Rainbird Publishers 125TL
Prof. R.J. Harrison-Church 74L, 102B, 122B, 123B
Harvard College 10, 46BR
Brian Hawkes 72T, 89T, 91TR, 127T C
Michael Holford 4C, 7C, 83T
Institute of Geological Sciences 113BR
Institute of Oceanographic Sciences 126TL
Sarah King 112C
Kitt Peak Observatory 4BL, 11C, 6B
Lowell Observatory 39B
The Mansell Collection 5BR, 20R, 21L, 22L, 24B, 25B, 26B, 27, 30LR, 115L, 120B, 124B
David Mallott 86T
Natural History Museum 69L
Novosti Press Agency 16TL TR, 52L R, 60, 63BL, 64B, 69L
NASA 36T B, 35BR, 40B, 43B, 48, 49BL BR, 50BL, 51T BL BR, 54TL TR, 53T BL, 55TR B, 56L R, 57BL
BR, 58T, 59R, 61C, 63BR
Parke and Roche Establishment 5T
Photri 3TB, 16BL, 17T, 23, 25TL TR, 26T CL C CR, 28T, 34T, 35T, 38T BL, 39T C, 40T, 42T, 45, 50BR,
53BR, 55TR, 57T, 59L, 61T B, 62T, 68T
Popperfoto 15, 17BL, 37L, 71R, 84B, 87B, 88BL, 94B, 107BL, 128B, 50C
Radio Times Hulton Picture Library 9BL 11BR, 33, 35BL, 37R, 38BR, 67BL, 71L
Walter Rawlings 81T, 100T, 127BL
G.R. Roberts 47T, 82T, 94T, 102T, 103T, 122T, 126T, 73L, 79B, 81BR, 82B, 83BL BR, 90B, 91B, 98B,
99L, 105BL, 110B, 112B, 125BR, 126B
Ann Ronan Picture Library 41
Royal Astronomical Society 21R, 22R, 29, 31B, 34B, 47C
Space Frontiers 117T
Science Museum 8B, 11BL, 18B, 32, 68B
Scala 7T
Shell International Petroleum Company 123T
Spectrum Colour Library 97C
Stanley Belicki International Press Service 64T, 62B
Suddeutscher Verlag 88BR
John Topham Picture Library 79TR, 87TR, 90T, 98T, 100C, 108T, 109TR, 4BR, 5BL, 9BR, 14, 16BR,
42B, 67BR, 77BL, 78, 92L R, 95L, 96, 101L, 103BL, 116, 111BR BL, 121BL, 127BR
John Watney 113TR
Zefa Picture Library 13, 18T, 24T, 44L, 58B, 87TL, 91T, 93T, 97T B, 101T, 103C, 105C BR, 106T B,
107T, 111T, 112T, 113BL, 118B, 124T